THE SCIENTIFIC ACHIEVEMENT
OF THE MIDDLE AGES

THE SCIENTIFIC ACHIEVEMENT
OF THE MIDDLE AGES

RICHARD C. DALES

UNIVERSITY OF PENNSYLVANIA PRESS
PHILADELPHIA

First Pennsylvania Paperback edition 1973

Library of Congress catalog card number: 73–77810

ISBN: (paperbound edition) 0–8122–1057–3
ISBN: (clothbound edition) 0–8122–7661–2

Printed in the United States of America

PREFACE

Men have tried an interesting variety of ways of dealing with the perceived or "natural" world in which they find themselves. They have imagined it as governed by hostile or benevolent whimsical forces, which they have tried to bribe or propitiate. They have explained it in terms of elaborate mythologies. They have tried to plumb its mysteries by interpreting flights of birds, innards of beasts, positions of the stars, or delirious mutterings of divinely inspired persons. They have tried to control it through the arts of magic and make it serve man's needs. Or, every now and then, they have sought to understand it according to the categories of human reason. It is such a rationally organized body of knowledge about the natural world which today is called science.

Science, as an organized body of thought, is dependent for its form on the culture in which it develops. Assumptions about nature, the world view and the value system all play a part in making up science, but they simply define the general directions which science may take. It is impossible to explain entirely how science is formed, or even why at certain periods, such as the Middle Ages, it has prevailed over less rational methods of thought.

This book is an attempt to provide an accurate sampling of medieval scientific thought in the context of a historical narrative, as well as a variety of evaluations of medieval science by modern scholars of many different points of view. It has been my aim to present as many actual medieval works as possible, but the diffuse and highly technical nature of some of these has often made it preferable to use competent summaries or paraphrases by leading modern scholars.

The field of medieval science is so vast that any attempt to be comprehensive in a book of this type was clearly out of the question. The selection is to some extent arbitrary and has been determined largely by my own knowledge and interests. It is also true that I have omitted those areas of scientific activity, such as medicine and biology, in which the medieval achievement was less notable. I hope that the book nevertheless accomplishes its purpose of

presenting to the modern student enough of the key works of
medieval science that he can make a reasonable judgment concerning
it and have some basis for discriminating among the various
opinions presented in the concluding section.

All translations are mine unless otherwise indicated. The book
is without footnotes; all editorial comments are placed in square
brackets in the body of the text. The notes have also been deleted
from the reprinted sections of other men's works.

Richard C. Dales

TABLE OF CONTENTS

LIST OF FIGURES

SCIENCE AND THE CULTURE
OF EARLY EUROPE

Adelard of Bath, an English scholar and linguist who had spent many years travelling, as he said, "for the sake of knowledge," returned to his native land in the year 1126. He had some particularly unkind things to say about the state of knowledge, especially that form which we would call scientific knowledge, in England and some particularly enthusiastic things to say about the state of scientific learning in other parts of the world that he had seen, places about which his countrymen knew very little. At the outset of his travels, which are quite remarkable in themselves for an Englishman of the early twelfth century, Adelard shared the diminished legacy of Latin antiquity in the West. He had been educated in a system which placed extraordinary emphasis upon literary, particularly scriptural, exegesis, possessed little or no knowledge of Greek or Arabic, and recognized no need for any but the most practical technological and mathematical skills. Within a few decades, practically within Adelard's own lifetime, both the educational system and the attitudes of Western Europeans toward the physical sciences and mathematics changed profoundly. In the century following Adelard's death, European thinkers were faced with the difficult task of assimilating the immeasurable and perplexing intellectual riches that, because of Adelard and others, they had so suddenly and unexpectedly acquired. By considering the circumstances of Adelard's life and unusual career, the traditions he learned and rejected, and the results of the reforms he proposed, we may perceive something of the first scientific revolution in modern European history and appreciate the enormous cultural shock that it represented.

I

Adelard was born at Bath probably around the year 1080, and studied at Tours and Laon, where it is known that he taught in the first decade of the twelfth century. Then, for reasons unknown, Adelard began to travel, to South Italy, Sicily, Greece, Cilicia,

Syria, Palestine, and probably Spain, returning to England in 1126, and probably resuming his travels during the 1130s. Adelard was in Bristol in 1140–1142, and there he met and dedicated his treatise on the astrolabe to the young Henry Plantagenet, later King Henry II of England. He probably spent the remaining few years of his life in England; it has been suggested in some capacity at the Exchequer, the royal accounting office, and he died around 1150.[1]

Adelard's criticism of the state of knowledge in England, produced shortly after his return in 1126, consists of three main points.

First, Adelard complained, his contemporaries were blinded by the prestige of ancient authorities and gave short shrift to modern thinkers. Second, the "Arab masters," whom Adelard greatly admired, had taught him more than he could ever have learned at home, and third, reason, not authority, must be the ultimate judge of learning. Why had Adelard travelled to the "Arab masters" to seek the knowledge he prized so highly? And what was the basis of his attack upon the Christian Latin dependence upon authority at the expense of reason? The answers to these questions not only tell us much about Adelard and his world, but they also constitute a general description of the place of scientific knowledge in the Latin West between the fifth and the twelfth centuries and of the cultural conditions that defined that place. Lest Adelard be made by his highly critical remarks to seem rather more sympathetic to recent attitudes towards science than he was, it is first necessary to observe that he did not mean by "reason" and "authority" quite what a modern reader might expect. It is also necessary to consider some of the serious results of his search for knowledge. Adelard's *Natural Questions* (see below, pp. 38–51), it must be said, is hardly a landmark in the history of scientific literature. Very little discussed in this work depends directly upon the "Arab masters" he praised so highly, and much of it consists precisely of those questions posed by the authorities of whom he was so scornful. The variety of questions Adelard discusses, however, and the ontology that many of them presuppose suggest graphically some of the distance that separates even the most "progressive" twelfth-century scientific investigator and his modern descendents. Adelard discusses, among many other topics, why plants are produced without the sowing of seed; why we hear echoes; why the fingers are of unequal length; why women, being more frigid than men, are more lascivious in their desires; if the sphere of the earth were opened, where a stone dropped into it would fall; what food the stars eat, if they are animals.

These samples of Adelard's work are as revealing as his critique of his countrymens' scientific knowledge. Alongside problems that any beginning student of botany, physics, biology, or meteorology still deals with today, there are other questions, equally "scientific" for Adelard, that no student of any modern science would dream of asking, or even wondering much about privately. For all his criticism of authority and his praise of "reason," Adelard, if he can be called a scientist at all, was a particular kind of scientist. To understand him fully, it will be necessary to ask some questions of our own, first about the character of scientific knowledge in the early twelfth century and the circumstances surrounding its acquisition, and second, about the place of that knowledge in Moslem and Christian culture between the seventh century and Adelard's return to England in the first quarter of the twelfth. The remaining parts of this essay will deal with those changes in the character and place of science in the Latin West that had already begun within Adelard's lifetime and continued through the thirteenth century, with the response to them, and with their place in the culture of early Europe.

First, the travels. Adelard did not get his scientific interests or knowledge in England or France, he had to travel quite far to acquire what he had, and his travels took him to the very edge of the Christian world and beyond—to Moslem Spain, Norman-Moslem Sicily, Syria, and Greece—and he at last found what he had been looking for in a cultural and religious world towards which most of his contemporary Europeans were inordinately hostile and about which they knew almost nothing.[2] Adelard was not the first European who had to travel great distances in order to find out about things he wanted to know, and although the necessity for travels as extensive as his was soon to diminish, he was certainly not the last. Three hundred years before Adelard, when Charlemagne wished to assemble a body of sufficiently learned men to put order and coherence into some of the fundamental texts of essential Christian knowledge, he had to ransack the whole of his vast dominions to find a dozen or so scholars who were qualified to begin the great task he set them. From Italy, Spain, Septimania, England, and Ireland scholars were invited to the emperor's court at Aachen. Most of the centers and schools they established, even with the support of Charlemagne and his successors, did not long survive, and the late ninth and tenth centuries witnessed once again, except for the existence of a few centers of learning where certain kinds of knowledge were preserved, the dispersal of its intellectual forces.[3]

In the tenth century, Gerbert, a young monk from the Burgundian monastery of Aurillac, went off to northern Spain to acquire some mathematical skills unobtainable at home, then travelled to Rome and to Reims to learn logic, also unobtainable at home. After a long career as a scholar and teacher (and tutor to a future king of France and to a future emperor), Gerbert was made pope in the year 999. Gerbert's career was productive, and he was in many ways a remarkable individual, yet Gerbert's learning was regarded by many as being so strange that it must have had diabolical origins, and the unsavory reputation of being a magician and a sorcerer hung about his name for centuries. Besides his studies in mathematics and logic and his reputation as a magician, Gerbert appears to have been a mechanical technician of no ordinary skill, and, as we shall see below, the development of such technical skills and the place of technology generally were to become increasingly important in the centuries to follow. Less than a century after Gerbert, Anselm, a young man from Aosta in northern Italy, left home in order to find a monastic retreat in which he might find a level of devotion and learning superior to those he knew at home. After missing a number of monasteries where he might have found what he was looking for, chiefly because he did not know of their existence, Anselm finally landed at Bec, a monastery recently patronized on a relatively lavish scale by the ambitious and energetic dukes of Normandy and only barely become by Anselm's lifetime a respectable center of devotion and learning, chiefly through the efforts of Lanfranc, another Italian wanderer, who was then its abbot. Sometimes such excessive journeys and hardships suffered in the name of devotion, learning, or both were rewarding. Gerbert became pope, and Anselm, the greatest logician between Plato and Descartes, became first, abbot of Bec and later, Archbishop of Canterbury. No aroma of dubiously acquired knowledge hung about the figure of Anselm, however, and his devotional austerity and holy life made him a saint as well as a scholar. More frequently, however, learning had to be its own reward, and the career of Adelard, far humbler than those of Gerbert and Anselm, was far more typical of the sociology of knowledge in the eleventh and early twelfth centuries.

The problems involved in finding out where learning was to be had and then taking the pains to go and get it loom large in early European intellectual history, and they constitute an important and frequently neglected dimension in the early history of science. For the few who wished or were able to make them, these travels were a commonplace, and they remained a commonplace through most of the twelfth century. Not until the period between the late twelfth and the fourteenth centuries did centers of learning grow up, sustain themselves for more than one or two generations, acquire a reasonably wide reputation, and attract teachers and pupils in any but the smallest of numbers. Yet travel and the ignorance of the whereabouts of centers for learning were only two of the social and cultural conditions that shaped early European thought. The ease and speed of communications, occasions for the general exchange of ideas, the preservation of books and records, social recognition and patronage, and the fertilizing activity of related disciplines and other areas of intellectual activity all attended the birth of "modern" science in the sixteenth and seventeenth centuries, and they still influence its activities. Few of these conditions were present for scholars and thinkers from the age of Charlemagne and before to that of Adelard and beyond. Books were few, expensive, difficult to read, located unpredictably in whatever center of learning or other ecclesiastical establishment they had happened to land, and not recorded in catalogues nor preserved in anything remotely resembling a decent library. Those who owned them often could not read them and did not know what they contained. The kinds of learning that were pursued at most of the ecclesiastical centers where books were kept was based upon a form of literary studies that rigorously subordinated logic, mathematics, and natural science to other kinds of knowledge. Teachers and their whereabouts were generally unknown and often located in remote and dangerous places. There was no reason for significant support of scientific studies, and the development of other disciplines more often than not retarded the growth of the sciences. Such retardation was not, of course, primarily the result of substantial hostility between such areas as theology and science, but rather that of the character of the educational process that produced ecclesiastical leaders, which, being based chiefly upon techniques of literary exegesis, regarded other forms of knowledge as ancillary to its main purpose. Considering the condition of knowledge in these other disciplines, such an attitude was not illogical.

Two other conditions should be considered in connection with the sociology of learning in general before the twelfth century. Technology, which established a particularly close and enormously productive relationship with theoretical science between the fifteenth and the eighteenth centuries, remained, with some exceptions, generally separate from scientific thought until well after Adelard's lifetime. Thus, not only did early European scientific culture suffer from the absence of such profound influences as technology has exerted upon more recent science, but it suffered also from the lack of that direction which science can give to technology. To be sure, as we have already noted in the career of Gerbert, that remarkable predilection for technological innovation that constitutes so essential a part of western cultural history was already at work in the tenth century. It enabled western Europeans to develop the most extensive variety of nonhuman sources of power and put them to more complicated uses than had any earlier civilization. As recent writers have shown, the influences of technology and, perhaps more important, technological con-conceptualization were felt in art and theology as much as in economy and industry, particularly in the twelfth and thirteenth centuries. Yet technology probably contributed little to the development of scientific thought in the tenth and eleventh centuries.[4] Moreover, those technological creations that have proved most useful to experimental science—accurate measuring devices, temperature controls, and closely machined instruments—did not appear until late. Man's ability to recreate certain sets of physical conditions repeatedly with absolute, unvarying regularity, the basis of experimental science, was thus hampered and further reinforced the widespread cultural belief that material creation seemed characterized by irregularity, mutability, decay, and intractability. As we shall see below, this attitude also had impeccable philosophical credentials as well as powerful theological support.

Besides the difficulties arising from the peculiar relationship between technology and theoretical science, there were epistemological difficulties as well. The very modern habit of conceptualizing bodies of knowledge as distinctive in methodology, vocabulary, and appropriateness for certain situations, that permits modern thinkers to speak of "scientific method" and even "science" itself, is a relatively recent human acquisition. Indeed, the process that allows the conceptualization of learned "disciplines" is itself one of the very foundations of modern knowledge. As anthropologists and cultural historians have repeatedly shown, not all societies nor even a single society in all periods of its history possess the ability—or the interest—to think in such terms. It is useful to remember as we investigate Adelard's "authorities" that his own language really possessed no word designating "nature," in the modern sense, and the terms that described knowledge were hierarchically arranged. *Scientia*, the Latin word from which the word "science" is derived, meant something like "practical information," and *sapientia*, "wisdom," was reserved for only one area of mental activity, the study of theology.[5]

So far, we have considered medieval science before Adelard's time primarily in terms of those cultural and social conditions that it shared with other forms of intellectual activity between the sixth and the twelfth centuries. Yet these were not alone in shaping the world of Adelard of Bath. Let us turn to Adelard's concept of knowledge and his critique of what he calls the "authorities" at the expense of what he calls "reason." The epistemological structures that shaped the character of education and the place of the sciences within the scheme of knowledge that education was designed to provide had not taken their form between the sixth and the twelfth centuries, nor was their origin particularly Christian. They had grown up in the cultural world of late antiquity, chiefly between the second and the fifth centuries A.D., and they reflect part of the final transformation of that world.[6] In Greek and Roman antiquity, from the Ionian philosophers to the astronomers of Alexandria, scientific speculation, mathematics, and medicine had been extremely fertile areas of intellectual activity. In the fourth century B.C. Aristotle alone had produced a vast and coherent compendium of studies in the natural sciences as well as in logical method and ethics, metaphysics, literary criticism, and political theory. The fullest development of the natural sciences in antiquity, however, was prevented by several conditions: the limitations of physical investigation, the classical scorn for manual labor, the enormous reputation of Aristotle, and the increasing interest in the consideration of ethics and metaphysics at the expense of the physical sciences.

Moreover, metaphysics, particularly as developed by the successors of Plato, acquired a strongly religious cast, and the vigorous spiritual consciousness of the Greek and Syrian East further contributed elements of religious thought, imbued with animist, spiritualist, and magical theories of causation. Although mathematics, astronomy, and medicine continued to flourish as late as the third and fourth centuries A.D., Platonic and Neo-Platonic metaphysics imposed a view of the material world that substantially altered earlier approaches to the study of the physical sciences. For the Neo-Platonists in particular, the Platonic theory that material reality was composed of inferior copies of pure ideas led to the conclusion that natural phenomena were both less interesting and less important than the ultimate types of which they were but feeble imitations. Pure ideas and the formless matter out of which things were made interested these thinkers, but not the individual instances of form being imposed upon matter. For them, mathematics alone, not the analysis and description of the physical world, led to the highest kinds of knowledge.

From the second century A.D. on, Neo-Platonism, with the religious cast of its thought, exerted a powerful influence upon both late paganism and Christianity, and it thus further reinforced the philosophical rejection of the study of the material world. The earliest Christian considerations of such topics, found in the literature of commentaries on the opening chapters of the book of *Genesis*, reflect precisely such Neo-Platonic influences.[7]

Not only did late antique philosophical and religious thought reject the study of the material world in favor of the consideration of spiritual reality, thus leaving science in a much humbler position than it had attained in the days of Aristotle and the later Alexandrians, but the means of preserving scientific knowledge, in literary form and in general education, also changed. From the second century B.C. there had begun to appear encyclopedic compilations and handbooks that tended to atrophy scientific knowledge and pass it on as unchallengeable fact with no methodology at all. The most characteristic representatives of this tradition in Rome, Seneca and Pliny in the first century A.D., reflect both the generally high level of popular interest in natural philosophy and the inconsistencies, gaps, errors, and absence of discipline that were inevitably preserved in works that were as much the product of leisured, curious gentlemen as repositories of exact scientific knowledge and methodology. Even the deficiencies of Seneca and Pliny, however, are of a much higher quality than the work of their successors. As the social and cultural world of Rome itself changed in the third and fourth centuries, the quality of the encyclopedic tradition declined. The conflicting views of the material world that had come down from earlier antiquity, the identification of areas in need of further investigation, and the downright mistakes that appeared with greater frequency all might have been resolved if the theory and practice of science (and technology) had occupied a different place in the values of late antique culture.

The immediate Christian successors of the late antique pagan compilers of handbooks and encyclopedias suffered both from the changes in late antique culture and from the limitations of their own interests. Increasingly, knowledge of many kinds came to be considered useful only insofar as it was deemed necessary for the proper understanding of Scripture and the writings of the Church Fathers. Although much has been made of the alleged hostility of early Christians to late pagan culture, its intensity and universality have often been exaggerated. Certainly much of the antique pagan legacy, including echoes of Seneca and Pliny, was preserved, although sometimes in rather surprising ways and in language whose imprecision and instability are striking.

St. Isidore of Seville, an early seventh-century theologian, compiled a work called the *Etymologies*, which became the greatest storehouse of antique learning for later Christian readers.[8] The *Etymologies*, in twenty books, became widely popular, and although they constitute an impressive (and interesting and often charming) intellectual achievement, they often preserved more errors than facts, and the following indictment of Isidore by a recent historian of medieval science will illustrate sharply Isidore's deficiencies:

Isidore has even less to say about geometry than arithmetic. Beginning with a strange, four-fold division of geometry into plane figures, numerical magnitude, rational magnitude, and solid figures, he concludes with definitions of point, line, circle, cube, cone, sphere, quadrilateral, and a few others. Here we find a cube defined as "a proper solid figure which is contained by length, breadth, and thickness," a definition applicable to any other solid (Euclid defines it as "a solid figure contained by six equal squares"). A quadrilateral figure is "a square in a plane which consists of four straight lines," thus equating all four-sided figures with squares![9]

In Isidore's case, we can see clearly the conceptual and linguistic shortcomings that attended the last stage of late antique encyclopedia production.

Changes in the literary history of scientific thought paralleled equally important changes in the structure of education during the same period. Late antique education had been primarily literary and rhetorical in its structure and had as its chief aim the training of the free (and leisured) citizen. Private and public conduct and the knowledge appropriate for an active civic life determined its subject-matter and the character of those to whom it was made available. The social changes of the third through the fifth centuries altered this model considerably. Fewer citizens participated in the public activities for which the old system had been designed, and those who did often came from remote parts of the Empire and rose to high office after military service with little intellectual preparation. It was partly to serve these people with the minimum amount of general information they could handle that the handbook tradition was further debased. The educational system, now patronized by the emperors, tended to take the form of a limited, structured number of subjects. These subjects, seven in number, became the liberal arts. Slowly over several centuries, they became divided into two distinct parts, the *trivium* and the *quadrivium*. The *trivium* was the basic form of education, and it contained grammar, rhetoric, and dialectic (logic). The *quadrivium* was a more advanced mathematical structure whose parts were arithmetic, music, geometry, and astronomy. The *trivium* remained the more important of the two, and for centuries it constituted the only education most literate people ever received. The *quadrivium* possessed a strong mathematical basis, and the increasing difficulty in finding people who could teach one or more of the subjects in it made it far more vulnerable to neglect. Although the *quadrivium* was still an important part of knowledge and learning as late as the seventh century and had been extraordinarily influenced by the work of the early sixth-century Roman philosopher Boethius, it was far less frequently taught between the seventh and the eleventh centuries. By the fourth century A.D., elaborate, ornate, and heavy-handed descriptions of the liberal arts had become a commonplace, but they were of little help in recovering the ground lost in science since the second century.[10]

This encyclopedic and educational tradition passed virtually intact to Christian barbarian Europe. Neo-Platonic metaphysics, with some adjustments, became one of the bases of Christian theology. The *trivium* and *quadrivium*, more usually the *trivium* alone, became the format of Christian education. Thus, two of the most important changes in the cultural world of late antiquity, the change in the structure of education and the transformation of the genres of philosophical expression from the treatise and the dialogue to the commentary on an authoritative—and, in many cases, sacred—text, shaped the next ten centuries of Christian learning. This, then was the origin of that system of authorities on such an unassailable level that resort to the writings of an accepted author alone often was sufficient to end an argument or prevent consideration of the questions he had dealt with. It was the impact of the reputation of Aristotle all over again, but with the canonization of far less original and competent thinkers than Aristotle.

In addition, the knowledge of the Greek language, in which much of the scientific learning of antiquity remained, gradually faded in the Latin-speaking West, and, as we shall see, the rise of Islam in the seventh and eighth centuries removed from the Christian West and the Imperial Roman East many of the territories in which such study was still vigorous. The language of these new conquerors, Arabic, later emerged as an important influence in the West in the time of Adelard of Bath.

The transformation of the place of scientific investigation and scientific thought in late pagan and early Christian culture, the powerfully religious application of such knowledge as had survived, and the influence of the new spiritual and intellectual framework of Latin culture contributed one further element in the existence of science, the creation of a cosmology whose influence in many aspects of European culture endured until the seventeenth century. In part, the construction of medieval cosmology was based upon the changes in the shape of knowledge that we have just considered. In part, it came about as the result of the relative unevenness of scientific knowledge and the influence of particularly distinct oriental and later Latin strains. Astronomy, having survived in rather better condition than other branches of science, came to determine the conception of the structure of the universe held by many thinkers. This conception was based upon the evidence of a geocentric universe, based upon the Aristotelian conception of the great difference between the composition of the material world between the earth and the moon's orbit on the one hand, and between the moon's orbit at the end of the finite universe on the other. Surrounding the universe was the realm of the fixed stars. Into this general and scientifically not improbable structure, there poured the creations of less developed science and theology—celestial intelligences, spirits, demons, and astral influences, all ranked according to an increasingly precise scheme, and all reflecting in macrocosm what individual human beings and their mutable world illustrated in microcosm—the blueprint of God's creation. The best short account of the history of this model cosmology and its widespread influence has been given by C.S. Lewis, and a further elaboration of it is not possible here.[11] Suffice it to say that in the early stages of the process of its formation, this cosmology exerted considerable influence upon both the direction and the substance of scientific theory. Ironically, its foundation was not bad science, but precisely the best and most precise science that the ancient world had produced.

So far, our discussion may have offered some suggestions as to the character of the authorities criticized by Adelard of Bath and other twelfth-century thinkers, how they had achieved their stature by the eleventh century, and how vulnerable they in fact were. Yet how did Adelard and others know of their vulnerability? Adelard's lifetime had witnessed no significant turn towards experimental science, and many of the questions Adelard considered in the *Natural Questions* had long ago been considered by Isidore of Seville and others. To answer this question, we must turn to two other topics which, although they were not originally related, together constituted one of the most powerful influences on twelfth- and thirteenth-century European thought: the rise of logic in the educational curriculum and the traditions of Islamic science.

Not only were the *trivium* and the *quadrivium* frequently separated and the latter often not taught at all between the seventh and the eleventh centuries, but within the *trivium* itself the study of grammar and rhetoric received far more attention than logic. This is partly the result of the emphasis that early medieval culture placed upon literary training for the study of sacred texts and partly the result of the character of materials available for the study of logic. Among the great legacies of the sixth-century Roman philosopher Boethius had been a plan to translate all of Plato and Aristotle into Latin. Of that plan very little had actually been accomplished, and it is useful to remember that until the late eleventh century all that the Latin-speaking world had of Plato was part of the dialogue called *Timaeus* accompanied by several much longer and obscure commentaries upon it, and all that it had of Aristotle was several introductory treatises on the study of logic, translated by Boethius, along with Boethius' own commentaries and his translation of Porphyry's *Introduction to the Categories of Aristotle*.[12] This slim body of introductory materials was not appealing to many teachers or scholars, yet it had before it a great future. In spite of its small quantity, it represented a monumental achievement. Not only did Boethius preserve Aristotle's introduction to logic, but in translating it and other works and composing his own commentaries on them he devised in Latin a vocabulary capable of expressing the subtle mental processes that had hitherto been discussed only in Greek. By systematizing a philosophical vocabulary in Latin that could accurately describe and analyze mental processes, classify valid arguments, detect logical errors, and precisely describe objects external to the mind, Boethius left later Europeans the tools, humble as they might have seemed, for the later elaboration not only of science and logic, but of law, theology, history, psychology, and metaphysics as well.

By the beginning of the tenth century, logic began to be considered once more with enthusiasm. The career of Gerbert, although it marks a distinct stage in the re-evaluation of logic, is only a preliminary stage, but within two centuries the study of logic had come to occupy practically the center of education, had already entered the consideration of theology and law, and had begun to attract teachers and students in great numbers. The story of the re-ordering of the *trivium* with the new emphasis upon logic has been brilliantly told by R. W. Southern.[13] The rise of logic generated in its train new applications of formal thought, caused new theological crises, and raised to prominence such thinkers as St. Anselm, Lanfranc, Berengar of Tours, and Roscellinus in the eleventh century and Peter Abelard in the twelfth. The influences of this intellectual revolution shaped the scholastic philosophy of the thirteenth century as well as the new restructuring of theology and law in the twelfth. It marks a genuine turning-point in the intellectual history of the West.

This revival and transformation of the study of logic occurred without the immediate addition or discovery of any works from classical antiquity. Shortly after it began, however, new literature did become known in the West, dealing not only with logic itself, but with the entire intellectual range of two cultures: that of ancient Greece and that of modern Islam. The revival of the study of logic predisposed western thinkers to treat the new discoveries of Greek and Arabic philosophy and science in new and extraordinarily productive ways, but it did not cause those discoveries. The great debt owed by Latin Christian and later European culture to the thought of the world of Islam is centered here, for the "Arab masters" cited so authoritatively by Adelard of Bath were indeed in a position to criticize the humble scientific and mathematical achievements of the West. They were the heirs of an important segment of the intellectual legacy of late antiquity, that vast corpus of medical and scientific knowledge that had been preserved in the Greek- and Syriac-speaking parts of the old Roman Empire, and barely at all in the Latin-speaking West. During the expansion of Islam in the seventh and eighth centuries, much of this knowledge and people who could study and teach it came under Arab rulership, and the old eastern centers of scientific learning influenced the eager, acquisitive culture of their new Arab masters:

Not direct contact with Byzantium, but the long-enduring Hellenism of the Syriac-speaking clergy of Mesopotamia fed the courtiers of Harun al-Rashid with translations of Plato, Aristotle, and Galen, just as they had formerly ministered to the curiosity of Khusro I Anoshirwan.[14]

Thus, classical philosophy and science was passed to the Arabs by the same Syriac-speaking Christian clergy that had earlier translated Greek learning for the Sasanid emperors of Persia.

The ninth and tenth centuries witnessed the translation of most of Greek scientific and philosophical literature into Arabic as well as the production of the first of the great Arabic commentaries on the philosophical and scientific writings of the Greeks and original Arabic discoveries in medicine, mathematics, astronomy, and philosophy. Ibn Sina (d. 1037, Avicenna to later Latin writers) wrote philosophical and medical works that acquired immense influence, one of them circulating in print in the West as late as the eighteenth century. Ibn Rushd (d. 1199, Averroes to later Latin writers) was perhaps the greatest of all commentators on Aristotle's philosophical writings. The work of these and other Arabic thinkers mark one of the greatest chapters in the history of philosophical and scientific thought. Not only the theoretical science of the ancients interested these men, but the possibilities inherent in it for further work, including the possibilities of experimental processes. Mathematical systems of the Greeks, such as Euclid's geometry, were joined by original Arabic contributions: algebra, trigonometry, and probably the invention of the zero. In these formal philosophical and scientific fields, as well as in architecture, music, agriculture, and poetry, the Islamic fertility of thought and expression was one of the richest in the world. From the eleventh century on, the West began to acquire this legacy for its own.

First, a small trickle of translations, chiefly from Spain, brought an intimation of what was available to a small number of thinkers in some parts of Europe. The process of translation was extremely awkward. First, the Arabic text had to be read aloud, then it was translated into Hebrew or Spanish and then retranslated into Latin, a process certainly not rapid nor free of error and misunderstanding. The possibilities for translations increased, however, when the expansion of Christian political conquests in Spain and Sicily and the Western Mediterranean in the late eleventh and early twelfth centuries brought many centers of Islamic learning into Christian hands. Perhaps the most striking example of the "instant" cultural consequences of this Christian expansion is to be seen in the case of Toledo. This city, long a center of Arabic learning, was captured by the Christians in 1085. In 1126, the year of Adelard's return to England, Toledo received a new Christian Archbishop, Raimund, who, within four years, began to develop a local school of translators from Arabic, of whom the greatest, Domenicus Gundesalvi, produced a large number of translations. Raimund was also responsible for the translation of Ibn Sina's works into Latin. During the same few years across the Mediterranean similar translations were undertaken. At the court of Roger II of Sicily, the great Islamic geographer al-Idrisi dedicated his most ambitious work to the king personally. At Salerno in South Italy, long a center for the most ambitious study of medicine in the Christian world, the traditions begun in the late eleventh century by such translators as Constantine the African and John the Saracen continued through the twelfth century. In 1127, the year after Adelard returned to England, Stephen of Antioch, an Italian, translated the *Al MaTaki*, one of the best Islamic medical encyclopedias. Adelard himself later translated Euclid's *Elements* from Arabic, and the greatest of the early translators, Gerard of Cremona (1114–1187), translated over ninety works from Arabic, including Ptolemy's *Almagest*, the standard textbook on astronomy in the West until the seventeenth century. In 1142 the Koran itself was translated into Latin under the sponsorship of Peter the Venerable, Abbot of Cluny. These translators, as well as their late twelfth and thirteenth-century successors who translated directly from Greek, thus provided Latin Europe with much of its forgotten Greek legacy as well as the extensions of that legacy produced by Arabic thought.[15]

II

The revival of the study of logic and its widespread application across the spectrum of twelfth-century thought and the acquisition of Latin versions of Greek and Arabic philosophical and scientific thought greatly invigorated the intellectual world of twelfth-century Christian Europe. Even at the very beginning of the great age of translations, western thinkers, strongly impressed with the initial results of the revolution in logic, had begun to draw up ambitions schemes of knowledge, several of which opened very real possibilities for a new epistemology.[16] The twelfth century also witnessed the first results of the application of the new logical studies to the diffuse and intractable materials of theology and law.[17] The work of Hugh of St. Victor, Peter Abelard, Gratian, and Peter Lombard represent some of the most important and consequential steps in this process. The new intellectual paths these thinkers followed, not, to be sure, without criticism, expanded and debated by the new society of the schools and the early universities, constituted the framework for the reception of the new translations from Arabic and Greek. The social dimension of this intellectual change, constituted by the existence and prominence of the new schools and the careers of those who passed through them, reveals a new social market for learning, and a new scale and kind of demand for learning and the careers learning might provide. Within this new world, the contributions of Islamic thought were quickly absorbed, not merely into the works of individual thinkers, but into whole academic curricula. The arts course at the thirteenth-century universities, which was the basic preparation for the advanced study of law and theology, consisted largely of materials in logic, mathematics, and the natural sciences.[18] So pronounced was the logical-"scientific" character of thirteenth-century academic curricula, that students of a more literary persuasion complained that logic and the sciences had killed off the humane letters.[19]

Indeed, the literary education that had formed the core of the *trivium* between the seventh and the twelfth centuries had by the thirteenth century become a preparatory course of study for the more advanced courses in logic and the natural sciences, while the highest ranges of academic study were reserved for those subjects that had themselves only been recently organized by the influence of logic in the eleventh and twelfth centuries, theology and law. Alongside this radical change in the curriculum, there had also developed characteristic forms for the academic presentation of the subject matter in all disciplines in the lectures and the disputations of the schools. In this atmosphere, thinkers arranged their new materials and presented them to eager, critical, and disputatious colleagues and students. During the course of the thirteenth century the application of logic to theological study that had so occupied the twelfth century was duplicated, this time by the attempt to fit Aristotelian philosophy and Arabic science into the broad, yet limited model of Christian cosmology and orthodox doctrines. Much of the history of the thirteenth-century universities swirls around Aristotle and the "Arab masters" and the objections raised against such scholars as Albertus Magnus and Thomas Aquinas, who worked to incorporate Aristotelian subject matter into the frame of Christian beliefs and thought. These great intellectual conflicts shaped the atmosphere surrounding the reception of Arabic and Greek thought in the twelfth and thirteenth centuries. These struggles not only raised issues that troubled western minds for centuries, but they also prepared the way for the late thirteenth-century assault on Aristotelian thought that continued until the seventeenth century. Indeed, the very revolt against Aristotle itself constituted another set of those circumstances that make up the background of the sixteenth- and seventeenth-century scientific revolution.[20]

III

In one respect, the recovery of Aristotelian natural science and philosophy and Arabic thought may well have contributed to and certainly was aided by another phenomenon of the twelfth- and thirteenth-century Christian West: the new place of sensory perception and the assessment of the place of material creation in a legitimate understanding of the operations of the universe. By introducing a legitimate means of investigating the material world and deriving accurate information from it, the Aristotelians of the thirteenth century finally found a philosophical rationale that could effectively attack the Platonic traditional Christian cosmology and metaphysics. Moreover, in doing so they developed a set of intellectual criteria that enabled others to criticize their own first steps toward establishing a genuine body of scientific knowledge. That is the most important legacy of Adelard of Bath and his successors. It would not, to be sure, be quite accurate to call them modern scientists, because they continued to work within a strictly limited Aristotelian frame of reference, and Aristotle, for all of his accomplishments, stood much improving upon. The achievement that enabled others to take the next logical step in this process, that of criticizing Aristotle on the basis of the new learning, formed the real threshold of the scientific revolution. As Lynn White has remarked,

We have been warned likewise that much of the earliest of this science must not be judged by modern notions of what science is about: deeply impregnated with Aristotelianism, it was often, at first, a qualitative science expounding a hierarchy of essences rather than a quantitative science discovering laws of mechanical efficient causation. The really important thing to be noted, however, is the rapidity with which the scientists of the later thirteenth and fourteenth centuries learned to differ with Aristotle, once they had understood him, and with which they created the new nonclassical fashion in science.[21]

These thinkers, whose scientific achievements are clearly open to criticism such as this, had nevertheless accomplished much. They had readjusted the traditional plans of knowledge and created important room in them for mathematics and physical science. Their successors, from the late thirteenth through the fifteenth centuries, used a language they had helped to create, a set of tried criteria for analysis and proof, and a new assessment of sense experience and the material world, even when they argued most vociferously against the increasingly mechanical and unthinking application of Aristotle's theories to all problems of natural science. These tools had their origins in the experience of those travellers, translators, students of logic, and cosmologists of the late eleventh and twelfth centuries and the "Arab masters" to whom so many of them had gone to school.

The conjunction of the appearance of new kinds of schools in the twelfth century and a new public with new purposes for learning and a new curriculum with strong emphasis upon logic and the natural sciences had important cultural and sociological consequences. The new directions of thought, however, and the characters and careers of those who followed them should not dominate our awareness of the revolutionary character of this period in other areas of thought, feeling, and expression. As historians of art have shown, the second half of the twelfth century witnessed a vigorous change in the representational treatment of material reality.[22] The great dialectic between Platonism and Aristotelianism that had gone on down through antiquity and the early middle ages had left its mark on art as well as science. Was the material world and its phenomena a mere illusion, an imperfect, irregular copy of a mathematically and spiritually superior immaterial reality? Or was it, on the other hand, as carefully made by God as the heavens themselves, was its operation regular and did it obey laws perceptible to human reason? Although the adoption of Aristotelianism gave a new philosophical role to the study of the material world, it was not able by itself to replace Platonic metaphysics by a "scientific method." It could not do so first, because not all of Aristotle's science was accurate, and because Aristotle's metaphysics was far less thorough and less attractive than Plato's; second, because some of Aristotle's and Avicenna's (and some of their Christian successors') conclusions concerning the eternity of the world, the collective soul, and the relations between substance and accidents in theological matters, among other propositions, seriously contradicted several of the most deeply-held dogmas of Christian theology. The revolution in the study of logic and the recovery of the Aristotelian corpus of knowledge had indeed had important consequences: they contributed to the change in both the conception and the organization of knowledge; they shaped the curriculum at the lower levels of university education and the minds of those scholars who proceeded on to the advanced levels of law and theology; they helped to restore the study of the material world, so neglected by Platonic metaphysics, to a place in the legitimate human study of divine creation; finally, they influenced the whole structure of the first relatively widespread educational system in modern history and in doing so brought to the common intellectual experience of a growing body of literate people the mental organization, discipline, and a possibility for intellectual growth that the earlier limited training in literary exegesis had not provided.

The consequences of these changes occurred in other areas besides new attitudes toward and interest in the study of the material world. Although the sources of the new realistic depiction of material reality in art—from human beings to flora and fauna—that appeared in the late twelfth century lie in other places as well, the new role of science also influenced the innovative extension of techniques in painting and sculpture. In the representational arts, in new styles of devotion that reflected a greater emphasis on both the human aspects of divine figures and the possibilities of individual spiritual growth, in the elevation of nature to a new honorific status, the change in natural science can be seen as part of a greater shift of sensibility that constitutes one of the watersheds of European culture.

Charles Homer Haskins once pointed out that "to distribute medals for modernity is not the task of the historian." It must be noted that not all of the changes discussed above led in a linear way to the "modern world" and to "modern values." In fact, few of them did. The first flowering of meticulously observed representation of nature in the visual arts did not endure, as Millard Meiss, among other historians of art, has so conclusively shown.[23] The spiritual—and later, the magical—powers long thought to be regularly and frequently operative in the material world never quite lost either their philosophical legitimacy or their fascination. The mental structures that made an appropriate and important place for these in early European ontology and cosmology seriously implied a criticism to man's ability to perceive more than an insignificant fraction of the forces at work in the universe. In the thirteenth century, even the meticulous physical observations contained in so remarkable a work as the Emperor Frederick II's treatise *On the Art of Hunting with Birds*, one of the most accurate and detailed accounts of falconry ever written, were less the result of homogeneous change throughout an intellectual system than the sudden—and quite discrete—spurt of interest in a small area of that system, and they do not herald anything resembling a scientific revolution.[24] C. S. Lewis once reminded his readers that new learning often brings with its new ignorance, and he illustrated his point dramatically by a brief analysis of the strangely dual careers and beliefs of important sixteenth-century thinkers, people who have often been regarded as the exclusive heralds of "modern" civilization but also can be shown to have believed wholeheartedly in systems of spiritualism, magic, witchcraft, and numerology.[25] To these qualifications could be added other examples. In the early seventeenth century, James Napier invented the system of calculation by logarithms, but he developed this system in order to calculate more efficiently the number of the beast in the Book of

Revelations. Among the earliest examples of mathematical probability theory, also in the early seventeenth century, were attempts to test according to mathematical principles the degree of veracity in the Gospel accounts of the life of Jesus. New awareness of, and new values placed upon material phenomena did not in themselves constitute a cultural revolution or utterly abolish an earlier system. It is both dangerous and inaccurate to single out one or two changes in a given historical system of thought and reward them for being "modern" while at the same time singling out other aspects of that system and condemn them for being "barbarous" or "backward." It is risky to isolate elements of culture from the matrices in which they have their existence and to establish a model according to which, for example, good modern science drives out bad older science in a kind of progressivist inversion of Gresham's law. One of the greatest values of a book such as this one is to afford the reader interested in the history of science and the student of cultural history access to an important and often neglected and misinterpreted body of knowledge that, far from being somehow different and better than the other elements of early European culture, constitutes one of its most important and characteristic dimensions.

IV

The documents chosen by Professor Dales to illustrate this dimension of early European culture range from the twelfth to the fourteenth century, precisely that period when so much other change engaged the minds of thinkers. Although most of the writers represented in these documents are usually considered chiefly in terms of their intellectual importance, it is useful to remember that the scientific professions as such did not exist, and most of these men had public carreers, much of their time and intellectual energies being taken up with the day to day business of the Church and the world. Robert Grosseteste, for example, was Chancellor of Oxford and later Bishop of Lincoln; Thomas Bradwardine was a great theologian and Archbishop of Canterbury; Nicholas Oresme was one of the most remarkable economic theoreticians of his age and died as Bishop of Lisieux. The texts in this book are then, for the most part, the work of thinkers deeply involved in the society and culture in which they lived. There are few "pure research" scientists here, nor is there anyone who felt particularly alienated, in the modern sense, from his own time and place. Of all the writers included here, perhaps Roger Bacon, mystic, theologian, and scientist, was the most isolated, although even Bacon's strange career and his subsequent reputation as a magician seem now to stem from his outspokenness within the Franciscan order rather than his unique scientific work.

The themes and topics Professor Dales has chosen to illustrate medieval science are those that show such thought at its most original—in the fields of natural science, the study of motion, and astronomy—at its most characteristic—the perplexing world of nonmaterial causation and the realm of magic that Dales calls "the fringes of science." Such fringes, as we have seen, were in part a legacy of late antiquity, and they survived well beyond the fifteenth century. Nor have modern physical and social science and modern politics been without their own unique "fringes": concepts of ethnic or racial superiority and inferiority, national destiny, the relation of the state to the individual. Such "fringes," of course, have not always been as clearly visible to those influenced by them as have the "fringes" of early European science, but they have played no less complete roles because of their relative invisibility, no less destructive or creative roles. Although the history of science and scientific conceptualization from the twelfth to the twentieth century often may seem a spectacular instance of the overcoming of those fringes it may be useful to remind ourselves that this history itself took the shape it did, at the outset, precisely because of the values of the culture out of which it grew. It was in the twelfth and thirteenth centuries that reason became regarded as a distinctive, unique, human possession that placed great responsibility upon those engaged in intellectual activity; it was the thinkers of the twelfth and thirteenth centuries who came to attribute to human reason and the material world that they discovered themselves so superbly equipped to investigate integral and honorific places in the divine plan of creation. These attitudes too are part of the complex legacy of early European culture out of which grew the scientific interests and the scientific work that this book so ably illustrates.

EDWARD PETERS

Philadelphia, 1973

NOTES

1. The best introduction to the lives and work of medieval and later scientists is usually the appropriate volume of *The Dictionary of Scientific Biography* (New York, 1970–). The following notes will be kept particularly brief because of Professor Dales' own Bibliographical Essay (below, pp. 177–182). This essay itself may be complemented by the thorough essay in Edward Grant, *Physical Science in the Middle Ages* (New York, 1971), pp. 91–115.

2. Generally, see R. W. Southern, *Western Views of Islam in the Middle Ages* (Cambridge, Mass., 1962).

3. Standard accounts are M. L. W. Laistner, *Thought and Letters in Western Europe, A.D. 500–900* (2nd ed. London, 1957), although this work is not particularly strong in the history of science; Philippe Wolff, *The Awakening of Europe* (Baltimore, 1968); R. R. Bolgar, *The Classical Heritage and its Beneficiaries* (New York, 1964), pp. 13–129.

4. On technology generally, with particular emphasis upon its place in medieval culture, see now Lynn T. White, Jr., "Cultural Climates and Technological Advance in the Middle Ages," *Viator* 2 (1971), pp. 171–202, with the works there cited.

5. For the general conditions of scientific revolutions, see Thomas S. Kuhn, *The Structure of Scientific Revolutions* (2nd ed. Chicago, 1970). For *scientia* and *sapientia*, see E. F. Rice, *The Renaissance Idea of Wisdom* (Cambridge, Mass., 1958).

6. The best recent study is Peter Brown, *The World of Late Antiquity* (New York, 1971), with brief bibliography.

7. See F. E. Robbins, *The Hexameral Literature: A Study of the Greek and Latin Commentaries on Genesis* (Chicago, 1912), and Lynn Thorndyke, *A History of Magic and Experimental Science* (New York, 1923), Vol. I.

8. On Isidore, see Laistner, Index, *s.v.* Isidore (St.). The best study is J. Fontaine, *Isidore de Seville et la culture classique dans l'Espagne wisigothique*, 2 vols. (Paris, 1959) and the most recent bibliographical study is J. N. Hillgarth, "The Position of Isidorian Studies: A Critical Review of the Literature since 1935," in *Isidoriana* (Leon, 1961), pp.11–74.

9. Grant, *Physical Science in the Middle Ages*, pp. 11–12.

10. H.-I. Marrou, *A History of Education in Antiquity* (New York,

1964); Pierre Riché, *Education et culture dans l'occident barbare, VIe–VIIIe siècles* (Paris, 1962).

11. C. S. Lewis, *The Discarded Image* (New York, 1964).

12. In general, see R. Klibansky, *The Continuity of the Platonic Tradition During the Middle Ages* (London, 1939) and the works cited in the next note.

13. R. W. Southern, *The Making of the Middle Ages* (New Haven, 1953). See also Bolgar, *The Classical Heritage.*

14. Brown, *The World of Late Antiquity*, p. 202. See R. Walzer, *Greek into Arabic* (Oxford, 1962)

15. On Islamic Science, see *The Cambridge History of Islam*, Vol. 2 (Cambridge, 1970) and Norman Daniel, *Islam and the West: The Making of an Image* (Edinburgh, 1962). For the translators, see C. H. Haskins, *The Renaissance of The Twelfth Century* (New York, 1957); Bolgar, *The Classical Heritage.*

16. See R. W. Southern, *Medieval Humanism and Other Studies* (Oxford, 1970), pp. 42–49; Bolgar, *The Classical Heritage*, pp. 232, 435.

17. Besides the works cited above, see M.-D. Chenu, *Nature, Man, and Society in the Twelfth Century*, trans. J. Taylor and L. K. Little (Chicago, 1968); S. Kuttner, *Harmony from Dissonance* (Latrobe, Pa., 1960).

18. Grant, *Physical Science in the Middle Ages*, p. 21; L. J. Daly, *The Medieval University* (New York, 1961).

19. See E. K. Rand, "The Classics in the Thirteenth Century," *Speculum* 4 (1929), pp. 249–269; L. J. Paetow, *The Arts Course at the Medieval University* (Champaign, Ill., 1910).

20. Excellent introductory bibliography in Grant, *Physical Science in the Middle Ages*, pp. 101–103.

21. Lynn T. White, Jr., "Natural Science and Naturalistic Art in the Middle Ages," *American Historical Review* 52 (1946/7), p. 422.

22. A good popular account is Henry Kraus, *The Living Theater of Medieval Art* (rep. Philadelphia, 1972).

23. *Painting in Florence and Siena after the Black Death* (New York, 1964).

24. See Kuhn, *The Structure of Scientific Revolutions.*

25. C. S. Lewis, "New Learning and New Ignorance," *Introduction to English Literature in the Sixteenth Century* (Oxford, 1954), pp. 1–65.

CHAPTER I

THE EARLY MIDDLE AGES

The Romans, however excellent they may have been in other departments of human activity, had never cared to make the intellectual effort necessary to comprehend fully the science of the Greeks, much less add to it. From the first century B.C. in the works Varro and Cicero, to the sixth century A.D. in the works of Boethius, Cassiodorus and Isidore of Seville, Latin authors compiled handbooks containing, among much else, simplified accounts, sometimes competent but often confused, of portions of Greek science. Besides this, some translations of Greek scientific treatises, especially medical works, were made during the sixth century, many of them at the instigation of Cassiodorus. This supplied the early medieval writers with a body of ideas and information from which to extract what they felt was pertinent to their needs. In addition to these works, the writings of some of the Church Fathers also contained much scientific information, usually itself derived from the handbooks but employed in a context consonant with the needs and interests of early medieval writers.

By far the most important and influential of these Latin writers for passing on a portion of the Greek scientific achievement to the early Middle Ages were two authors, Calcidius and Boethius, writing nearly two centuries apart. Throughout the Middle Ages, their works were read by nearly all educated men, and between them they largely formed the context of pre-thirteenth-century Latin science.

The earlier of these two writers was Calcidius, who in the early fourth century translated into Latin the first two-thirds (including the story of the lost continent of Atlantis) of Plato's creation myth, the *Timaeus*, and wrote a lengthy commentary on it, based on several of the better Greek handbooks.

Plato had taught in his *Timaeus* that the artisan God, or Demiurge, had fashioned the cosmos from pre-existing unformed matter or chaos, using as his model the eternal Forms or Ideas. He described the creation of the elements by motion, the mathematical proportions that bound them together, and the structure of the universe.

His description of the elements employs a peculiar kind of stereo-metric atomism. The original creation from chaos was the scalene triangle. These triangles were then combined into a variety of solid forms, whose properties were determined by their shapes. The basic elements were earthy (solids), watery (liquids), airy (gases), and fiery (volatile substances), and there could be an indefinite number of "isotopes" of each, thus accounting for the variety to be found within each of the general classifications; in addition, mixtures of the four elemental types existed in most things.

His conception was highly poetic, his approach mathematical, and his means of communication often figurative and obscure. It was this enigmatic form of expression which made it necessary for commentators to "explain" Plato's meaning. The astronomical and cosmological sections particularly are obscure and capable of multifarious interpretations. Of the numerous *Timaeus* commentaries written between the third century B.C. and the twentieth century A.D., that of Calcidius was by far the most widely read and influential.

Whether or not Calcidius was a Christian, he translated many Greek words, phrases, and concepts in such a way as to alter the original meaning and make the *Timaeus* seem much closer to Christianity than it was. He also had a habit of using a phrase of the *Timaeus*—often taken out of context—as an excuse for a disquisition of his own, so that his commentary passes along much later Greek thought. There are traces of Aristotle (his description of the soul, the notions of prime matter and the transmutation of the elements, and Aristotle's cosmological scheme of concentric spheres), of the Stoics (the periodic recreation of the universe, and the concepts of natural law and "seminal reasons"), and a great deal of the Neopythagoreans and later Platonists. Calcidius's main source for astronomical information was the *Manual of Mathematical Knowledge Useful for an Understanding of Plato* of Theon of Smyrna, a Greek handbook author of the second century B.C. From this source he gave proofs for the sphericity of the earth and its location in the exact center of the universe, and he asserted that the earth is a mere point compared to the size of the universe. He then went on to discuss the celestial and terrestrial circles and the zones of the earth, the order, orbits and periods of the planets, the causes of eclipses, and the size and distances of earth, moon and sun. He also passed along the semi-heliocentric theory of Heraclides of Pontus, although he, like his source Theon, seems to have misunderstood it. It is known from other sources that Heraclides had held that the interior planets, Mercury and Venus, revolved around the sun, while the sun revolved around the earth; and that the earth rotated daily on its axis. Calcidius attributes an epicyclic theory to Heraclides, whereby the center of the epicycles for the sun, Mercury and Venus is the same, although the epicyclic theory was not devised until long after the time of Heraclides.

But whatever Calcidius's shortcomings, he provided the Latin Middle Ages with their only direct knowledge of a work of Plato, and he was much better equipped intellectually than most Latins to understand the complexities of highly sophisticated Greek technical works and to make these intelligible in Latin through his *Timaeus* commentary, one of the most able and fruitful works to come down to the early Middle Ages from late Antiquity.

The most competent of all Roman scientific writers was Boethius (*ca.* 480–524), of noble Roman family who served the Gothic king Theoderic in many important civilian capacities before falling from favor and being executed for treason in his forty-fourth year. Boethius was fortunate enough to have been raised during the period of relative stability and recovery of Theoderic's reign. He not only received the usual rhetorical education, but he also was well versed in philosophy and the sciences, and he had a good command of the Greek language. Realizing the sad state of intellectual culture in his day and fully aware of his own gifts, he determined to carry out a vast educational program for his Latin-speaking countrymen by translating or adapting the basic Greek works on the four mathematical subjects of the curriculum (arithmetic, geometry, music, and astronomy, the "fourfold path to wisdom," or Quadrivium, as Boethius named it), by providing Latin versions of the basic logical works, and finally by translating all the works of Plato and Aristotle and harmonizing the thought of these two great philosophers.

At the time of his death, only about half of this grandiose project had been completed, and not all of that has survived. But it is still an impressive achievement. More than any other Latin author, Boethius had a comprehension of the sciences and a realization of the need to read basic texts rather than encyclopedic compilations. Of his books on the Quadrivium, *Astronomy* (probably based on Ptolemy's *Almagest*) and *Geometry* (a free translation of Euclid's *Elements*) have been lost. *Arithmetic* is a free translation with some additional material and some omissions, of one of the better Greek arithmetic texts, that of Nicomachus. His *Music* is based closely on the Greek treatises of Nicomachus and Ptolemy and treats music from a theoretical and mathematical standpoint. In the field of logic, he translated two of Aristotle's elementary logical treatises, *Categories* and *Interpretation*, as well as commentaries on these works, and he may have translated the advanced logical works and *Physics*, although this is not certain. He never found the time to translate the major philosophical works of Aristotle or Plato or to write his synthesis of their teaching. His philosophical ideas, both Platonic and Aristotelian, are best expressed in his treatise *The Consolation of Philosophy*, a beautifully written work of mature philosophical reflection composed while he was in prison awaiting execution.

From Boethius's textbooks, the Middle Ages received their knowledge of some of the best ancient ideas on the mathematical structure of the universe, of mathematical proportions, of music theory and its relation to the harmony of the cosmos, of the types and limits of human knowledge, and of the classification of the sciences. They learned, too, of the wave theory of sound, the relation between frequency of vibration and musical pitch, and the definition of an atom. But most important, they learned to conceive of the world of nature as an ordered whole and to deal with it rationally. The techniques for rational treatment they learned from his translations, with commentaries, of the logical works. And from his *Consolation* they learned the use to which a ripe philosophical wisdom might be put. During the early Middle Ages, Boethius's works and translations were virtually the sum total of available information on arithmetic and logic and a very large proportion of what was known about music. Even during the most mature period of medieval science, the twelfth through fourteenth centuries, Boethius's writings were studied closely and with profit.

During the century after Boethius's death, the shaky stability of the Germanic kingdoms on western Roman territory was shattered by the wars of reconquest under Justinian and by the general ineptitude of the barbarian rulers. The fabric of ancient society, including its political, social, economic and educational institutions, fell apart. Literacy itself, let alone scientific advance, was in danger of disappearing. But as new social forms acquired some integrity, new values replaced the old, and new political arrangements came into being, the intellectual heritage of the ancient world was pillaged and reconstituted to meet new conditions, those of a frontier society of farmers, warriors, and missionaries of God's word. The highest level of culture probably still existed in ravaged and supine Italy, but the vigor and initiative now came from northern Europe.

Of several attempts to reconstitute the antique scientific corpus to meet the needs of a different culture, that of the north English monk Bede stands out. Bede's goals were all theologically determined, but he felt the necessity of an accurate knowledge of the world of nature in his theological pursuits. His principal scientific works were *De natura rerum (On the Physical Universe)*, *Liber de temporibus (On Times)*, and *De temporum ratione (On the Reckoning of Times)*. Although Bede drew his knowledge of the natural world from standard late antique sources such as Saints Ambrose, Augustine, Basil, and Gregory the Great, his *De natura rerum* was derived mainly from St. Isidore's work of the same name, itself a complication. It was, however, superior to that work for several reasons. For one thing, Bede had access to Pliny's *Natural History*, or an abridgment of that work, probably the best of the Latin scientific encyclopedias, whereas Isidore had not known it. More important though were Bede's clarity of thought, common sense, and assumption of natural causation. These qualities often permitted him to rise above the level of his sources. Isidore had been confused and inconsistent about the shape of the earth. Bede, on the other hand, held it to be a sphere at rest in the center of the universe and surrounded by the seven "heavens." The sphere of the stars revolved in a circle about the earth while the various planets followed epicyclical orbits around it. He understood enough to be able to give a clear account of the reasons for eclipses and for the phases of the moon. The corporeal world, in his conception, was separated from the spiritual by the waters of the firmament and was made up of the four elements, fire, air, water, and earth, whose combinations constituted the world of corporeal nature. The earth itself was divided into five zones: two frigid, one torrid, and two temperate. Both temperate zones were inhabitable but only the northern was actually inhabited.

He was particularly interested in problems of chronology and wrote two works on this subject. The first was a short, admirably clear presentation of the basic units of time and their relationships, called *De temporibus*. This work was an attempt to bring order out of the chronological chaos of the ecclesiastical calendar, the most pressing problem of which was the establishment of the date of Easter and other movable feast days. Since the date of Easter depends on the lunar calendar of the Hebrews, its date fluctuates in terms of the solar calendar of the Romans. A lunation and a solar year are incommensurable, thus making it necessary to adjust the two periodically. The Irish had developed a system of establishing the date of Easter, but it had many faults and was different from the Roman method. Aside from the Irish astronomical tables, called *computi*, Isidore was the major authority on questions of chronology, but he too left much to be desired. Bede undertook to explain clearly the major divisions of time and to point out the problems caused by their incommensurability. He also introduced his readers to problems of historical chronology and general problems of the measurement of time, and he explained the use of the nineteen-year cycle (by which nineteen solar years are considered equal to 235 months) in figuring the dates on which Easter would fall in the future. Although the *De temporibus* was an elementary work, it is admirable in the clarity with which it grasps and presents the factors involved in the problem and proposes well-thought-out solutions.

In 725 Bede published a much fuller and more comprehensive work on chronology, the *De temporum ratione*, which included many related subjects, such as computation, astronomy, comparative chronology, history, and geography. In this work he used a large number of antique writers on chronology as well as many chronicles employing different systems of dating, which he endeavored to synchronize according to a single chronological system. He adopted from Dionysius Exiguus, a bilingual writer on legal topics and a friend of Cassiodorus, the method of dating forward and backward from the Incarnation. Then he computed the major dates of world history on this scale and appended to his work an outline history of the world from the Creation to 725 A.D. Included in this work was a very important study of the tides, which combines experience with knowledge gained from literary sources. From Pliny, Bede learned of the correlation between the tides and the motion of the moon. But he also knew that the tides follow the moon by different intervals at different places, the interval remaining fixed for each port (this is the important principle known as "establishment of port"); and how the wind can retard or advance the tide. And he explains how the nineteen-year cycle can be used to predict the tides for each port.

A second, more far-reaching and comprehensive revival took place on the continent under the Carolingian dynasty. Between the beginning of the eighth century and the end of the ninth, by the heroic efforts of many men, a new Europe was constructed of many disparate elements. Not the least important aspect of this reconstruction was a new curriculum and a new educational system. Science, in our sense, was not an important part of the new curriculum, but it was present. Some basic knowledge about the structure of the world, enough practical astronomy to fill the needs of a Christian society, a knowledge of how to perform arithmetic computations, and a modicum of medical knowledge (consisting of extracts from the handbook authors, fragments of the Hippocratic corpus which had earlier been translated into Latin, and native empirical medicine and knowledge of herbs) were included in the curricula of most of the schools of Latin Europe. As the ninth century drew to a close, several Europeans wrote learned and competent commentaries on the astronomical portions of antique handbooks, but aside from these any scientific progress is most difficult to detect.

The tenth century, though on the surface a time of invasion, cruelty, barbarism and chaos, is nevertheless the turning point in European intellectual history in general and the history of science in particular. Contact with the Greek world, never completely broken off, began to increase, and the earliest known intellectual contacts with Muslim learning occurred near the century's end. A more fruitful use was made of ancient sources—always with a practical end in view—and the Europeans even began to add to their inherited store of knowledge. The monastic school at St. Gall provided an admirable education in astronomy (the monks even performed some observations), medicine and arithmetic.

The great figure of the tenth century was Gerbert. He was born an Aquitanian peasant, educated at a local monastery, taken to the Spanish March by the Count of Barcelona to complete his education, made tutor to the son of the Holy Roman Emperor, and was eventually elected pope as Sylvester II. His happiest and most fruitful years, however, were spent as schoolmaster of the cathedral school at Reims. Here he made great efforts to procure the best available textbooks in each subject, including a *Geometry* probably written by Boethius, which had not been heard of since the sixth century, and a work on *Astronomy* attributed to Boethius. The latter work is wholly unknown aside from Gerbert's reference to it. He himself wrote textbooks on the mathematical disciplines, including a work on the abacus. His curriculum also included medicine. But it was in his teaching of astronomy that Gerbert was most revolutionary and most successful. Most of the standard handbook authors were available to him, as well as the *Astronomy* which he attributed to Boethius. These authorities, however, present a variety of astronomical hypotheses, some assuming eccentric orbits for the planets, some utilizing epicycles, one (that of Aristarchus) placing the sun in the center, another (that of Heraclides of Pontus) assuming that only the interior planets, Mercury and Venus, revolved around the sun, while the sun and the other planets revolved around the earth. Gerbert, although fully aware of the inconsistencies of his authorities, consciously rejected the more sophisticated explanations and adopted the astronomy of Pliny, who admitted eccentric orbits but not epicycles, probably on the pragmatic grounds of simplicity.

In order to communicate the principles of astronomy to his students, Gerbert made two models, which he used as teaching aids. The first of these was a large globe, made of polished wood covered with horsehide, on which he drew a map of the heavens. The various circles, such as the horizon, the ecliptic (that is, the sun's apparent yearly path around the earth), and the celestial equator (the extension of the earth's equator to the sphere of the fixed stars), and the constellations were marked in different colors to aid in their identification. By means of it he was able to demonstrate visually to his students the effect of the obliquity of the ecliptic (that is, the angle at which the plane of the sun's apparent path is inclined to the plane of the celestial equator) in changing the rising and setting times of the constellations located on the zodiac.

An even more spectacular aid was his planetarium. Its basis was a system of intersecting metal rings, including one which represented the ecliptic. At the center was the earth. Along the ecliptic he placed the planets, probably represented by small metal balls, and by an ingenious system of wires he was able to move each planet in its proper orbit.

Gerbert was also a patient astronomical observer and constructed at least two fairly complex viewing instruments: a hollow wooden hemisphere, large enough to accomodate a man's head, with the circles of the five zones marked on it and six-inch viewing tubes of equal bore inserted at the intersections of these circles with a longitudinal circle; and a single viewing tube oriented to the heavens by means of two metal rings and outlines of the major constellations made of wire, by means of which, writes his pupil Richer, "if one constellation should be pointed out to anyone, even though he were ignorant of the subject he could locate the others without a teacher."

During the eleventh and twelfth centuries, Europe underwent a period of rapid growth and development in every aspect of her life. Population increased dramatically, trade and commerce revived, old towns grew and new ones sprang up. The pressures generated as a result of this growth caused far-reaching changes in the structure of European institutions. The progress of the sciences was at first less pronounced that that of many other elements of the culture, but after a slow start it developed considerable momentum. The process of translation of Greek and Muslim works, barely begun in the tenth century, proceeded apace during the eleventh and increased even more rapidly in the twelfth century. The main centers of translation were Spain, especially Toledo; Sicily; southern Italy, particularly Montecassino and Salerno; and several of the northern Italian towns, whose commercial and diplomatic relations with Constantinople created a small class of Italians with a literary knowledge of the Greek language. The works translated included both Greek and Muslim treatises on theology, philosophy, astronomy, astrology, mechanics, chemistry, optics, mathematics, zoology, geography, psychology—in fact, a considerable portion of the ancient scientific corpus, ranging from Aristotle in the fourth century B.C. to Ptolemy and Galen in the second A.D.—but very little in the way of pure literature. It was a rich heritage, acquired over a relatively short time, and it required some time to be adequately assimilated.

CHAPTER 2

THE TWELFTH CENTURY

The twelfth century was a transitional period in the history of medieval science. The old standard sources (especially Calcidius' translation of Plato's *Timaeus* and his commentary on that work, the various treatises on Boethius, Macrobius' *Saturnalia* and Seneca's *Natural Questions*) continued to dominate and the newly acquired works were often grossly misunderstood. The striking thing about this century is the attitude of its scientists. These men are daring, original, inventive, skeptical of traditional authorities although sometimes overly impressed by new ones, and above all steadfastly determined to discover purely rational explanations of natural phenomena. These explanations are usually painfully *ad hoc*, indicating the lack of a coherent, self-consistent world view or a carefully worked out method of investigation.

Nevertheless, the conviction of these scientists that nature operated in a uniform manner, according to rational laws which man had the power to discover, their intense interest in the natural world for its own sake, their habits of precise observation of natural phenomena, and the high value they placed on man as a rational being, portend a new age in the history of scientific thought. The two selections which follow are typical of the best of that time and should illustrate adequately the character of science in the twelfth century.

Adelard of Bath is one of the crucial figures of the transitional period of the twelfth century, when the Latin classics and the late antique handbooks and translations were still the dominant sources of scientific information and attitudes, but when the new translations from Greek and Arabic were just becoming known to European scientists. Adelard himself was a pioneer in this movement. Born in England during the last quarter of the eleventh century, and for much of his life connected with the English court, he went to France for his higher education, first studying at Tours and then teaching at Laon. Then, determined to perfect himself in the wisdom of the Arabs, he spent seven years in travel, visiting Italy and

Sicily, Syria, Palestine, and perhaps Spain. He became an ardent exponent of Arabic learning and did much to popularize it in Latin Europe as well as translate several treatises from Arabic into Latin. His works include a dialogue called *On Unity and Otherness*, in which he tells of his commitment to a life of philosophy as against the love of things of this world; the *Natural Questions*, selections from which are presented below; a work on the abacus; a translation of Al-Kwarismi's astronomical tables; a textbook on the four mathematical subjects; and a translation from Arabic of Euclid's *Elements of Geometry*.

The date of the *Natural Questions* has never been established with certainty, but it was probably written fairly early in Adelard's career. He had already taught for some years at Laon and had spent seven years in travel to familiarize himself, so he says, with the wisdom of the Arabs. However, despite his constant references to the Arabs in the *Questions* and his praise of their learning, the editor of this work has been able to trace only a little Arabic influence. Adelard's sources for the most part are the same ones which Europeans had been using since late Antiquity, and uppermost among them is Calcidius' translation of and commentary on Plato's *Timaeus*. His quotation of Aristotle (Ch. LXXIV) is from an intermediate source which has not been identified.

Adelard exemplifies the growing curiosity and intellectual adventuresomeness of European scholars in general during the twelfth century. His attitudes and values are new and different, but they are the causes rather than the results of the translating activity which occupied so much of the time and energy of twelfth-century scholars.

Selections from Adelard of Bath, *Natural Questions*, translated from the Latin text of Martin Müller, *Die Quaestiones Naturales des Adelardus von Bath* (Münster i. W., 1934). *Beiträge zur Geschichte der Philosophie und Theologie des Mittelalters*, XXXI. 2.

I. *Nephew*: Let us begin our disputation by talking about plants whose roots are in the earth. And so I ask what is the reason that plants arise from the earth. Since the surface of the earth is at first level and immobile, what is there that moves from it, that rises up, grows, and puts out branches? For since, if you wish, you may collect earth which is dry as dust and place it finely sifted in a vessel of crockery or brass; and then after some time has passed, when you see the plants rise from it, to what would you attribute this if not to the wonderful operation of the wonderful divine will?

Adelard: Certainly it is the will of the Creator that plants should rise from the earth. But this thing is not without a reason. In order that this may be completely clear, I concede that plants rise from earth. Not from pure earth, however, but from a certain mixture which contains all four elements with their qualities in each of its particles, which indeed can be perceived by our senses. These four simple substances make up the one body of the world in such a way that although these components exist in every separate compound, they are nevertheless not perceived by the senses. But we wrongly call the compound by the name of the simple element which predominates in it. For no one has ever touched earth or water, no one has ever seen air or fire. These composites which we perceive by sense are not the elements themselves, but are things composed of them. Therefore, they should not be called earth, water, air and fire, as the Philosopher [Plato, *Timaeus*, 49ᴰ] says, but earthy, watery, airy, and fiery; yet each is named for that element which is most abundant in it. Therefore, since all four causes would exist in that earth of yours, even though it is finely powdered, a certain compound necessarily arises from it which is mostly earthy, somewhat watery, less airy, and least fiery. This compound sticks together because of its earth, spreads out because of its water, and rises upward because of its air and fire. For nothing is able to move upward unless it has some fire in it. Also, unless it had water and air, it could not spread out sideways. And finally unless it contained earth, it would not stick together. Wherefore, those things which had been concealed by wonderful subtlety in your dust have come out into the open. However, in order that you might understand this clearly, I place the cause of this process in the fact that the exterior elements arouse and draw out what is similar to themselves, and by their qualities they call forth the same qualities. Therefore these inferior elements are perpetually dissolving into things similar to themselves. ...

IV. *Nephew*: But you shall not pass on to the next subject without some difficulty. You have been making a chain by which you shall be shackled. For if, as you say, all four elements are present in each composite body and that they can thus furnish nourishment to other composites, then the air which we see would suffice to furnish the same nourishment to plants, and therefore they would receive nourishment from it after they were pulled out of the earth. But, because air cannot furnish this to them, despite the fact that they desire air, your whole explanation is destroyed, and the accomplishment of all things must rather be ascribed to God.

Adelard: I take nothing away from God, for whatever exists is from Him and because of Him. But the natural order does not exist confusedly and without rational arrangement, and human reason should be listened to concerning those things it treats of. But when it completely fails, then the matter should be referred to God. Therefore, since we have not yet completely lost the use of our minds, let us return to reason. Certainly air, as well as earth, contains in itself the four seminal elements, but yet it neither can nor ought to furnish nourishment to an uprooted plant. See that it is as I say: Although every sensible object (by this I mean everything which is perceived by sense), has been put together from these causal principles, nevertheless each separate component is not present to the same degree in each composite body. For in this body there is more earth, in that water predominates, this has a greater share of air, that is aflame with fire. Therefore each individual thing follows the properties of those specific elements which it has the most of, and each thing is especially nourished by that element which predominates in it. Therefore plants, as you surely must know, have a greater share of earthy nature and therefore are especially nourished by earth. If they are uprooted, they do indeed find nourishment in the air, but the wrong kind. For they will get from it less of what they need much, and much of what they need less. For this reason an uprooted plant must necessarily be dissolved because, since the earthy part does not find sufficient nourishment in in the air, it wishes to be freed from this compound so that it may return to its like, namely earth. When the earth is removed, the other components are freed from its weight and can return to their likes. The common herd, which always seems to lack the proper name for things, calls this dissolution death, although it ought to be called not death but change. Whence the Philosopher [Plato, *Timaeus*, 33^C], speaking about the world, says: "For nothing went out or came into it from anywhere, since there was nothing; it was designed to feed itself on its own waste." And indeed in my judgment nothing dies altogether in the sensible world, nor is it less today than when it was created. For if any part is set free from one compound, it does not perish but passes on to another....

VI. *Adelard*: It is difficult for me to talk with you about animals, for I have learned one thing, under the guidance of reason, from Arabic teachers; but you, captivated by a show of authority, are led around by a halter. For what should we call authority but a halter? Indeed, just as brute animals are led about by a halter wherever you please, and are not told where or why, but see the rope by which they are held and follow it alone, thus the authority of writers leads many of you, caught and bound by animal-like credulity, into danger. Whence some men, usurping the name of authority for themselves, have employed great license in writing, to such an extent that they do not hesitate to present the false as true to such animal-like men. For why not fill up sheets of paper, and why not write on the back too, when you usually have such readers today who require no rational explanation and put their trust only in the ancient name of a title? For they do not understand that reason has been given to each person so that he might discern the true from the false, using reason as the chief judge. For if reason were not the universal judge, it would have been given to each of us in vain. It would be sufficient that it were given to one (or a few at most), and the rest would be content with their authority and decisions. Further, those very people who are called authorities only secured the trust of their successors because they followed reason; and whoever is ignorant of reason or ignores it is deservedly considered to be blind. I will cut short this discussion of the fact that in my judgment authority should be avoided. But I do assert this, that first we ought to seek the reason for anything, and then if we find an authority it may be added. Authority alone cannot make a philosopher believe anything, nor should it be adduced for this purpose....

XII. *Nephew*: Your discussion still leaves some doubts in my mind about the animals. I wonder, for example, why some of them can see more clearly at night than in the light, for we know that light is an aid to vision, and darkness is the contrary.

Adelard: In the eye, which is the instrument of vision, there must necessarily be several humors. The white one aids vision, and through it the visible spirit comes forth and is diffused. There is another one, which is black, which prevents the white one from being diffused too widely. Because it is dark, it compresses the white humor and holds it together. Therefore, those animals which see better by night than by day have a large amount of the white humor and very little of the black. And so in the daytime too much of this humor goes forth and becomes so diffuse that it loses the ability to distinguish the objects of sight. But at night, because of the exterior darkness, it is contracted into itself and goes forth more uniformly and distinguishes things more clearly. This point can be illustrated another way. You have seen some people who have eyes whiter than they ought to be, and so do not see very well. If you were to take them out to a broad plain white with snow, they would be able to distinguish little or nothing. And you yourself must squint when you wish to look straight at something so that the compressed sight does not wander but reaches the objects of sight with the power to discriminate....

XIV–XV. *Nephew*: One should listen to what you say but not believe it. But I shall gird myself for higher things, so that, as far as my little knowledge permits, light might come forth from the smoke. For although I am ignorant of the Greeks' boasts, and I have not seen Vulcan's cave [i.e., Mt. Aetna], nevertheless I have learned both to know what is true and to disprove what is false, and I have considerable skill in this. So continue! I want to find out what you think about human nature. For although you may consider what you have already said to be very important, nevertheless, if you do not know yourself, I think that your remarks have little value. For men ought most properly to investigate man. ...I want to learn about the composition of man. So solve this problem for me first: Why don't men have horns?

Adelard: In order to establish that your question is worth answering, you must first bring forth some true or likely reason why it seems they ought to have them. Otherwise such a question does not merit discussion among philosophers. For I am not one who thinks we should struggle to find the causes of all the things that exist, but only of those things which seem to reason that they should be otherwise than they are.

Nephew: That is a reasonable demand. Here is my explanation: Everything which the Creator brought forth from formless nothing into the form of being, just as it was made by the Best, so, reason shows us, was made in the best possible way. The Creator gave to all the things He made the capability of existing, and they have a strong desire to exist. And so that they might preserve their existence they were given the means to defend themselves. Some animals (and it is these I am talking about) have weapons which are part of their substances—for example, the boar has tusks, the lion claws, the bull horns—by which they can defend themselves from any danger which threatens them. But other animals, which are smaller, are unable to defend themselves because of their slight size, and these save themselves by swiftness of flight—such as goats, deer, hare, and others of this kind. Now I ask you why, when the lower animals have inborn means of defense, man, who is more worthy than all the rest of nature, is not born with any weapons, such as horns or lethal tusks, and cannot even avoid a threatening enemy by swiftness of flight. For this reason, since he has not been given any weapons by the Creator, he must laboriously provide for himself the weapons he needs, and when it is necessary he must put his trust in the feet of another animal rather than in his own. Therefore, that creature which is dearest to the Highest Goodness is destitute of the highest aid.

Adelard: First I will give the popular answer. For I believe that man is dearer to the Creator than all the other animals. Nevertheless it does not happen that he is born with natural weapons or is suited for swift flight. But he has something which is much better and more worthy, reason I mean, by which he so far excels the brutes that by means of it he can tame them, put bits in their mouths, and train them to perform various tasks. You see, therefore, by how much the gift of reason excels bodily defenses. ...

Indeed, man is a rational, and therefore a social animal, particularly suited for two types of activity, namely action and deliberation, which some people like to call war and peace. His day by day experience teaches him that the use of weapons is required in warlike activities, but truth teaches him to lay these aside in time of peace and to remove them far from his thoughts. Indeed, anger stirs up one of these, and reason calms the other. And so if man had natural weapons, he would not be able to put them aside when he was making a treaty of peace. Also, if his defense lay in swiftness of flight, he would lose his ability to stand firm, and in time of war he would find himself weak because of the unsuitable gracefulness of his limbs. Now, however, when it is necessary he takes up arms, and he lays them aside in accordance with the demands of peace. And when he needs to, he can run swiftly by putting forth a great deal of effort, and when the necessity ceases he is able once again to stand firm. ...

XVIII. *Nephew*: Since we have been discussing things having to do with the brain, explain, if you can, how the philosophers determined the physical location of imagination, reason and memory. For both Aristotle in the *Physics* [an erroneous reference] and other philosophers in other works, have been able to determine that the operations of imagination are carried on in the front part of the brain, reason in the middle, and memory in the back, and so they have given these three areas the names imaginative, rational and memorial. But by what skill were they able to determine the site of each operation of the mind and to assign to each small area of the brain its proper function, since these operations cannot be perceived by any sense?

Adelard: To one who does not understand, everything seems impossible: but when things are understood, everything becomes clear. I would guess that whoever first undertook this task learned something about it from sense experience. Probably, someone who had formerly had a very active imagination suffered an injury to the front of his head and afterwards no longer possessed the imaginative faculty, although his reason and memory remained unaffected. And when this happened it was noticed by the philosopher. And similarly injuries to other parts of the head impeded other functions of the mind, so that it could be established with certainty which areas of the brain controlled which mental functions, especially since in some men these areas are marked by very fine lines. Therefore, from evidence of this sort, which could be perceived by the senses, an insensible and intellectual operation of the mind has been made clear. For even though the mind itself is an incorporeal being and cannot be perceived by any of our senses, nevertheless, because of the sensible operations it performs in the body, we cannot doubt its presence there. For from the motion which the body is seen to have, and which it could not give itself, the existence of something incorporeal in the body is proved. ...

XXIII. *Adelard*: Since many men have expressed many views concerning sight, it seems fitting in this case that we should set forth each of them in turn and then determine which of them is most reasonable.

Nephew: Then if you approve, I will state each of the principal views, and you raise objections to them when I have finished.

Adelard: That is a good way to proceed.

Nephew: The opinions concerning sight which I have been able to collect from various sources fall into four categories. Some men say that the mind, situated principally in the brain and looking at the outside world through open windows, namely the eyes, grasps the forms of things and makes judgments about them, but in such a way that nothing goes out from it to exterior objects and none of the forms reaches it from the outside.

Others assert that sight results from forms approaching the eye, saying that the forms of bodies impress themselves upon the intervening air right up to the eye, and thus they pass on to the mind, which has the power to make judgments about them.

Then there are quite a few who claim that some sort of visible spirit is sent forth from the mind and that the forms of visible objects encounter it in mid-air. Then, having taken on the forms of these visible objects, the spirit returns to its seat and makes these forms known to the mind, which then makes judgments about them.

A fourth group is firmly of the opinion that no forms approach the eye from the objects of sight, but rather a certain thing, which they call a fiery force, first arises in the brain, then goes through the concave nerves to the eyes, and then all the way to the visible objects. It takes on their forms as if they were a seal [and it were wax], and by returning to its origin with the same swiftness, it carries these back to the mind. [Adelard and his nephew proceed to discuss each view at length, finally dismissing them all.]

Adelard: Finally, there is a consideration about which there is no doubt, which, even if the other objections are set aside, can overthrow all of the above-mentioned opinions, namely that we often look at our own reflections in the mirror. And this fact, although its truth is attested by daily experience, does not accord with what has been said above. Therefore, it would be fitting for us to set forth that view which the Philosopher [i.e., Plato] approves and, spurning the others as untenable, we should place our faith in this Academic truth. This opinion which I have been talking about is as follows: In the brain there arises a certain very subtle air, having a fiery nature and therefore extremely light, which is sent forth by the mind through the nerves to exterior objects of sight whenever it wishes or it is necessary. This is called "visible spirit" by physicists. This spirit, because it is a body, needs a place from which to exit. It finds an egress through the two concave nerves which the Greeks call "optic," extending from the brain to the eyes. Just so, going out to the body, it makes its way with wonderful speed to the object of sight, and having been impressed by this object, it receives and retains its form and, returning to its origin it makes this form known. Now this spirit is called "fiery force" by the Philosopher. This force, when it finds a mirror or some other reflecting body opposite itself, bounces back from it toward its own face. It then acquires the form of its own face and as it re-enters through the eyes it makes this form known to the mind. But you must not think that the fiery force found the form of the face in the mirror. Rather, in being bounced back from the surface of the mirror, which was too smooth for it to stick to, it picked up the form on its way back. And this is the divine opinion of the members of the Academy which Plato has asserted, among other things, in the *Timaeus* [*Timaeus*, 45-46]. ...

XXXI. *Nephew*: Now that this is all clear to me, I should like to hear about the other senses.

Adelard: I think I should keep my remarks brief on the other senses, especially since they proceed from much the same cause and are susceptible of the same explanation, which I shall now expound. Smell, taste and touch arise when the appropriate organs of each of these senses come into contact with exterior objects. Smell comes about when air which has been affected by the object of smell comes into contact with nipple-like organs which are connected to the brain and are the appropriate organs of this sense. Things to be tasted are brought into contact with the tongue and palate, which are the organs of taste. Likewise, the hand is applied to some object of touch, while others are brought into contact with the hand. Whence it is necessary that the qualities of these sense organs be altered in the following manner: [The organs perceive inequalities between themselves and the objects of sense, but in perceiving the inequality the organ is itself altered to conform to what is has sensed.] ...

XXXIII. *Nephew*: Now I shall propose for discussion something that happens to us as a matter of daily experience: What is the explanation for the fact that one can blow either hot air or cold from the same mouth, as he pleases: For since, as the physicists teach, all such air comes from one and the same lung and follows the same path, how is it that it can be made to have opposite effects so quickly?

Adelard: Since this can properly be doubted, we ought to hunt for the cause. It is well known, as you say, that both hot and cold air proceed from the lung, but this does not happen at random. For when air proceeds from the lung, if it is emitted at once through the open mouth, since it comes from a hot place and nothing happens to it on the way, it feels hot. But if it is not emitted at once but is held in the mouth, then it cools off somewhat because the mouth is not so hot as either the lung or the passage-way to the mouth. Then, if you wish to use it for cooling something, you can change it to cold by blowing it hard through your pursed lips; for no element is more easily changed into different qualities than air....

XLVIII–XLIX. *Nephew*: Why is it that the earth, which supports all heavy things (I am speaking now not of simple elements, but of compounds), stays in the same place? By what is it supported? For if all these heavy things, such as rocks and wood and other such things, need something to support them and cannot keep the same position in the air because of their weight, how much more would the globe of the earth, which is the heaviest thing of all, need things to support it and be unable to be held in place by the surrounding air. Therefore, it is not reasonable that it should remain still.

Adelard: But it does not fall; and in order that we should not fall with it, we shall show that it is reasonable that the earth remains stationary. We know that the primary quality of earth is weight. A thing which is heavy gets along better in the very lowest place, for each thing loves whatever preserves its life, and it tends toward that which it loves. Therefore, every earthy thing must necessarily tend toward the lowest of all places. But it is clear that in a round figure, the center and the lowest point are the same. Therefore, any earthy thing tends toward the center, which is a simple and indivisible point. Therefore, it is clear that all earthy things tend toward a definitely located simple point, and there is only one such center, not several, and it must therefore be occupied by one earthy body, not by several. Nevertheless, all the other earthy bodies, as we have said, tend toward this same point. Therefore one presses against another as each one hurries toward the same point. Each heavy body both hastens and falls toward the same point, for falling is nothing but the hastening of a heavy body to the center. But that central point toward which they are falling remains fixed. Therefore, while these heavy bodies are falling toward a conditon of stability, they do not cease to be stationary unless some other force acts on them which turns them aside from their natural tendency. And you will now not doubt that what you earlier thought to be the cause of falling, is the very same thing which gives stability and cohesion to heavy bodies.

They remain motionless because the point toward which they tend remains motionless.

Nephew: If a hole were made all the way through the earth so that there would be a clear passage from one sky to the other, what would happen to a stone thrown into it?

Adelard: The same cause which makes the earth as a whole stationary would also bring the falling stone to rest.

Nephew: You have said enough, for I understand that it would come to rest in the center....

LVIII. *Nephew*: I still have a question about the nature of waters. You remember when, some time ago, we spent several days at the house of a witch in order to study incantations. She had a certain vessel of wonderful power which was brought out at meal times. It had many holes in the bottom and the top was filled with water for washing our hands. As long as the servant stopped up the upper holes with his thumb, no water came out of the lower ones. But when he took his thumb off the upper holes, the water immediately came flowing out of the lower ones. I, thinking this was magic, said, "It's no wonder the old lady is a demonic enchantress, when her servant can do things like this." But you, according to your custom, since you are very much interested in enchantments, were less willing to attribute it to magic. What do you now think about that water? The lower holes were always open and yet nothing flowed out except at the will of the water-carrier.

Adelard: If it was magic, it was nature's, rather than any power of the water-carrier. For since the four elements make up this natural world, and they are joined together by natural love in such a way that no one of them wishes to exist without the others, no place either is or can be empty of them. Whence it happens that immediately when one of them gives up its place, another occupies it without any time intervening. And it cannot give up its place unless another, which it regards with a certain substantial love, is able to succeed to it. Therefore, when the succession of the second element is prevented, the exit for the water to run out will be opened in vain. And so, because of this love, this waiting, you will vainly open the exit for the water if you don't provide an entrance for the air....

LXXIV. *Adelard*: Next we must take note of the action of the stars, concerning which we must accept not my opinion, but Aristotle's (which is my view because it is his): "Whatever moves," he says, "must be moved by nature, by force, or by will. That which is moved by nature either moves upward, like fire, or downward, like earth. But the stars do not move in this way. Therefore, they are not moved by nature. They are not moved by force, either, for what force would be great enough? Therefore they must be moved voluntarily by their own will. And if they move by their will, it follows that they move as living beings." [Not directly from Aristotle, but based on *Physics*, VIII, 4–5; *On the Heaven*, IV, 3; *Metaphysics*, XII, 8.]

I shall now establish this beyond any possible doubt. If things are sometimes moved and sometimes stand still, and when they are in motion sometimes alter their course either forward or backward, they are not moved by nature. For whatever nature does, neither ceases nor changes. And so the only possibilities which remain are that they are moved by force or by will. But there is no force in nature more powerful than the turning of the outermost heavenly sphere, and they cannot be moved by this, for they move in the opposite direction. Therefore, they are moved by will. You see that this follows from what I have just said. Furthermore, if their action is the cause of life and death of the lower animals, what must we think about them? Indeed, when the sun recedes from Cancer and from our region, first the greenness of the plants withers, then the joy of the flowers bows it head, and soon not even the beauty of the leaves remains to the trees. Finally, it is well-known that there are many animals, both of the earth and the sky, which die in the winter and come back to life in the summer. So what must we say? Only a mocking jester would claim that something which gives life to other things is itself devoid of life. Also, whatever observes a definite numerical law and constant order in its motion, certainly employs reason. And nothing is more definite, nothing more stable in its order, than the course and numerical law of the stars; for when have the bounds of the zodiac been exceeded by the planets? When the sun touches the limit of Capricorn, it changes its course and comes back to us, but it only comes as far as the limit of Cancer. And other stars obey other rules. Among creatures there is nothing more reasonable than these. By the immortal gods, I beg you, I implore you, use your mind. Since those things which just awhile ago had rushed off toward Capricorn, then returned from it without any force being exerted on them, they either know that they should return from it, or they do not know. If they do not know they should return from it, why don't they keep going past it? Why do they return so precisely? But if they do know this, they have been endowed with knowledge. But they cannot have knowledge unless they have a mind of some kind. Therefore, they are alive and possess both knowledge and reason.

The following anonymous essay on the elements was written probably between 1150 and 1175 in southern Italy. It is a remarkable document in many ways. Its author, although depending mainly upon the traditional authorities, has produced a strikingly novel treatment of his subject. His world view is derived principally from Plato's *Timaeus* with Calcidius' commentary, but this is compounded with mechanistic atomism, a semi-Heraclitean view of fire, and bits and pieces of scientific lore culled from Virgil, Lucan, Seneca, Macrobius, Boethius and others, all this slightly leavened by a knowledge of Aristotle's *Physics*. These citations of the *Physics*, although they betray a most faulty understanding of Aristotle, are very likely the earliest verifiable direct citations of that work in Latin Europe. The author's confident assertions that motion and the world are eternal, that everything is composed of atoms and the void, and that the earth might possibly revolve on its own axis, are evidence of the freedom of the atmosphere in which he worked and of the daring and original nature of twelfth-century physical thought. The logical coherence and self-consistency of the essay are unusual for the time; although the explanations of natural phenomena still have a strongly *ad hoc* quality about them, they are at least consistent with the author's first principles and with each other. The author's method is still extremely crude—the invention of *ad hoc* mechanisms suggested or illustrated by rough analogies to everyday experience, but the quality of his thought is of a very high order.

This essay is one of the best examples of twelfth-century European scientific thought before the recovery of the major Greek mathematical and scientific works and the translations of Arabic treatises and commentaries had exercised any appreciable effect but were beginning to be known and to be taken account of. Within two generations after this treatise was written, the most important writings of the Greek and Muslim scientists became the common property of the European intellectual community, and this treatise on the elements must have seemed quaint to a mid-thirteenth-century scholastic scientist. Still, it was several centuries before what we choose to call the "scientific attitude" would be so clearly in evidence as it is in this scientific opuscule of a transitional age.

Selections from Anonymous, *On the Elements*, Translated and abridged from the text of R. C. Dales, "Anonymi *De Elementis*," *Isis*, LVI (1965), 174–189.

I

Nature (*physis*) properly speaking is a principle of motion which a thing has as a result of its own nature, and not as the result of some concommitant attribute. The elements fully qualify as natural according to this definition. Now let us briefly see what is the cause of the elements, and then investigate their powers and capabilities so that, having learned the origin and powers of those things from which every visible being on earth and in the sky came to be, we might better understand how all these things arose from the elements.

The origin of the elements is motion, and this motion has existed eternally... It cannot be otherwise. ...Two things cause motion, lightness and heaviness. Quickness depends upon heaviness (weight), as can be seen in the way a man walks and in many other things; for when a man walks he does not drag first one foot and then the other after him. In the beginning, the motion coming from the Craftsman immediately made two extreme and contrary elements at exactly the same time, namely fire and earth. ... And thus through motion the elements were made, as Plato says near the end of our Latin version of the *Timaeus*, mentioning the similarity of this process to the threshing of grain. From these elements come all the principles of things whatsoever. (Now certain physicists have claimed that the principle of all things is a single one of these elements, *i.e.*, fire, earth, air, or water. But the elements properly so-called are those things which have a principle of motion from themselves, and not from matter or from some other thing.)

Quickness of motion makes that which is lightest, namely fire, all of whose parts are in motion; and slowness makes that which is heavy. These two extreme elements mutually mitigate each other's properties: the quickness and lightness of fire are less because fire rests on something solid and is constrained by nothing; but the solidity and hardness of earth are less because the lightness of fire, which cannot exist without motion, perpetually compresses, by its motion, that which is slow and makes it solid and hard. (A hard thing is something whose parts cannot easily be cut.) And so these two things, quickness and slowness, go hand in hand and hold themselves ready to receive a subject. The other elements are placed between these two contrary extremes, one less light and one less heavy. Why there are two rather than one is the concern of arithmetic, not of physics. The one which is less light is called air, and that which is less heavy is called water. Air is not so light as fire, since it is only light because of a constriction of its nature.... The parts of fire are more continuous than the parts of air, those of air more than the parts of water, those of water more than earth—I do not mean "continuous" in the manner of a solid object, because the parts of the higher elements yield more easily to cutting. If water penetrates anything, it penetrates earth, because heavy things are more porous. Air penetrates water and earth; fire penetrates air and the others. Thus the origin of the elements is motion which, descending eternally into matter, makes all the elements.

II

Now that we have seen the origin of the elements, let us move on to consider their powers. Let us first investigate the powers of the higher elements, because the higher elements act, while the lower ones are acted upon. The two lower ones behave like matter, the higher ones like an artisan.

Let us first consider fire, which behaves like an artisan and acts upon the other three elements. Fire is fast-moving, subtle, and sharp or penetrating. It is hot, illuminating and light. It is fast-moving because right down to its simplest parts—that is, its atoms—it is in motion as a whole, and each one of its parts is moved by itself. As a result of so great a motion, it becomes subtle or sharp and penetrates all things; nor can it be contained in a vessel the way earth or water, or even air, can be....It illuminates because it makes airy, matter much lighter and confers quickness upon it. This quickness continues all the way to the point of our sight, and an illumination occurs in the air (for air alone is luminous and can be made so by the sun or a lamp). Just so, on the contrary, darkness is a slowness of the air, obscuring the quickness of the pupil of the eyes. Fire is hot because, by its power, motion so penetrates any material, watery or earthy, that it moves and divides each of the parts of such matter and makes each part move rapidly; and the outside of the solid body begins to burn, because heat is the power of fire that divides solid bodies. The other powers of fire all derive from this one. Because it moves rapidly, it penetrates, heats and illuminates.

It is also light; that is, by its own nature it tends to move upward. And whatever else tends to rise, does so as a result of the natural fire which is mixed with it. The reason that fire is light is that it is not surrounded and compressed by anything, and thus each of its parts is always in motion *per se*, and from this arises its amazing lightness. Constriction, on the other hand, brings about the opposite. Solidity makes slowness, because in solid objects the particles stick together in such a way that one does not move without the others, and so not just one particle, but the whole object, is moved. And if a stone is thrown, its particles do not change their place, but the ones near the circumference remain there, and the ones near the center remain near the center. But fire, because each part moves by itself, ...is thus extremely subtle and more continuous than the other elements. And because fire itself is compressed by nothing, it is the lightest of all things as a result of its motion. ...Anything which is utterly light requires no support, but heavy things do need a support of some kind. Fire, which is lightest, supports air; air supports water; water supports earth. ...Whence it is said that: Thou has established the earth upon the waters. ...

One ought to note what Aristotle says in the *Physics* [correctly *Topics*, V, 5] that there are three kinds of fire: a glowing coal, flame, and light. He is speaking here of earthly fire, which comes about in this way: The air is set in motion (that is, its separate particles are moved) by some means, such as a steady blowing or by striking hard rocks together, ...and it passes over into fire and and penetrates some solid object, that is, some earthy or watery matter, whose separate parts it moves and divides. And thus from this violent and wonderful motion, heat and redness appear and the object begins to glow. (I shall not explain at this point why it is redness rather than some other color, but I shall rather discuss that question in my treatment of colors.) To proceed: This motion extracts moisture from the solid matter and expels it in the air that penetrates the moisture; and because each of the separate particles is moved, the object becomes hot and flame arises. But just because the moisture is thus penetrated by motion, I do not therefore say that water is flame, lest I should speak like those unskilled physicists who say that in fire or in any other element, there is water and all other elements. But air is thinned out by the very great motion of fire and is made lighter, and here and there it ignites and thus light arises. But fire is invisible by its own nature, and what we see is not fire, but solid material thus occupied and covered over by the motion of fire. And so from this wonderful motion descending into some solid earthy or watery matter, heat arises, because, as we have said above, heat is the power of fire divisive of solidity. Whence it is that there is never heat in the air unless it is there as a concommitant attribute, that is, because of a cloud or of some sort of moisture, such as often occurs in this moist air of ours. Air, because of a certain degree of slowness that it possesses, both nourishes our substances and takes something from them. That it takes less from them when we stay at the proper elevation is clear, because there is a certain kind of air, such as on the peaks of mountains and other high places, in which, if we were to remain for three hours or some such time, we should die.

There still remains one power of fire, namely that it is dry. Now a thing can be called dry because it does not cause dampness, as is said of earth; or it can be called dry because it dries things out. Fire is dry in the latter sense. It penetrates something, such as a moist body, and by its motion it separates the particles of that body and causes it to pass over into the nature of air or fire; and thus it dries a thing out by taking the moisture away from it, and by its motion drawing the moisture into its own nature or that of air. For example, if water were poured all over a table, and fire were placed near it, that moisture would gradually disappear and the table would become dry. Nevertheless, the moisture is not annihilated. It is only that its parts have been separated by the rapid motion of fire, and it has passed over into airy matter. For this reason, where there is more moisture, there is more smoke. For example, in a cauldron over a fire, because of the motion of heat, the particles of moisture are separated and certain tiny round drops rise in the air and become clouds; and then the smoke is less, because smoke is nothing but a large number of very tiny drops transforming themselves into air. Whence, in summertime clouds which are seen high in the air quickly dissolve because they are separated by the motion of the nearby heat, and they pass over into the nature of air or fire, and thus they seem to be annihilated. But where there is less heat, these drops run into each other and become larger and are borne to earth; and this is what causes rain. But we just mention this in passing. In what follows it will become more clear when we inquire into the causes of hail and rain, winds, thunder, and similar things.

Now that we have seen the powers of fire, let us pass on to the powers of earth so that, having learned the powers of the extremes, we might better understand those of the means. Earth has each of its powers exactly contrary to the individual powers of fire. Fire is fast-moving, earth is immobile—that is, it does not move with respect to its parts; just as, on the contrary, fire does move with respect to its parts, and one part impels another, and they do not stick to each other. With earth it is just the opposite, because it does not move with respect to its parts, but one part sticks to another so that if one moves, they all move. The physicists, such as Macrobius, call this dullness (*stupor*). Aristotle's statement in the *Physics* that the earth is immobile must be understood in this sense, that is, it remains still by its own nature and its parts are not moved. For no one denies but that it may well revolve on its own axis....Nevertheless, this will not now be said in physics, because physics requires probability, and this does not seem probable to all. Another way it can be known that the earth remains still by its own nature is that, since our senses, such as sight, and the other senses, employ a high degree of quickness, it is necessary that some solid and stable matter resist them in order to be known, and so only earthy and watery matter can be sensed.

Also, fire is subtle, while earth, on the contrary, is obtuse, that is, corpulent or solid. And as was the case with fire, we can make three distinctions of this one power. Earth is obtuse, that is, it does not penetrate. If earthy matter ever does penetrate, it does so accidentally, that is, as a result of a violent impulse supplied by something else. The reason that earth itself does not penetrate is its immobility and slowness, and the conjunction and sticking together of its particles.

Also, earth does not illuminate but rather makes shadows—and so in the fables it is called the seat of the dead and the shades—because darkness is a slowness of the air resulting from the absence of the sun or any other light. Earthy matter causes a shadow in the direction opposite where the light is coming from, because the ray of the sun reaches the solid matter which, because of the slowness and immobility of its parts, blunts and weighs down the air and diminishes its quickness. Therefore, the air in the opposite direction cannot be illuminated, and so there is a shadow. And so the shadow is measured by the magnitude of the earth's mass. We will not now explain what a ray of the sun is, but will go into this later. But it is true that the air is slower and heavier in dark places than in others, such as caves and underground places, where men of phlegmatic and humid constitutions are quickly weakened and do not last long. But "dry" men can last longer, although they are

more quickly consumed in high places. But we shall say more about
these things later.

Earth is also the seat of coldness. You can see the cause of cold
by blowing hard or gently on your palm. If the breath strikes the
palm with great force, it clogs up the pores: the pores extend through
the whole flesh and skin, and through them sweat and other super-
fluities are ejected. But these pores are blocked and clogged up by
a violent impulse of the breath, and thus the breath, which is
naturally hot, feels cold. But if it reaches the palm gently, it feels
hot, as is its nature. Similarly, from an impulsion of earth or colli-
sion of air, as occurs as a result of winds, water is compressed, and
by this great compression it is forced over into the earthy nature as
into ice or rock: and gems such as crystal and other shiny stones
come from water. But we shall speak of this later.

Also, earth is heavy and slow, because its nature is to weigh things
down. This is because earth makes things heavier by its slowness and
complete lack of sharpness and forces them over to its own nature.

Also, earth is dry. It is called dry because it does not moisten,
whence it is also called arid. Fire is called dry in a different sense,
because it dries things out. ...Fire dries them out by consuming and
dividing moisture; earth does so by compressing them and by unit-
ing with their particles, since it is the property of earth to compress.

...Now that we have seen the powers of the extremes, let us
speak of the powers of the means, which have some things in
common with the other elements and certain things quite different.

Let us first discuss air. Air is light by its own nature, that is, it
does not offer any resistance to the senses (unless perhaps through
a concommitant attribute). A thing is properly called light because
it offers no resistance to any sense. Don't we feel the air in our faces
when the wind blows? This is not as a result of its own nature,
but rather of a concommitant attribute, that is, from earthy matter
mixed in it and from a violent impulse, as has been said. Also,
doesn't the air look green [!] when we see it in its pure state? This
is sheer illusion caused by a ray of the eyes, of determined extension,
which becomes weak and imagines itself to see something when in
fact it sees nothing. ...But some people say that there exists a
certain crystalline body, and that when there are no clouds it is

seen purely. This is completely false. Rather, a ray of the eye which is directed upward in the morning imagines this because of a defect of vision; just as, when we close our eyes, we think we see shadows when in fact we see nothing, and we assert the opposite of what is obvious, because from the absence of the senses we dream up many things. ...Isn't the air seen just as clouds are? This is the result of a concommitant attribute: because of the slowness of some moisture, that which was closely packed together is thinned out. Again, isn't air seen when flame runs upward into the air? This too is the result of accidental moisture which lights it up. Doesn't the air offer some resistance to the sense of hearing? Yes, but it is because of an accidental impulsion that all these things are sensed in the air, and not from its own nature. Concerning these and other mutations of the air, which are called "passions" of the air by physicists, we shall speak fully when we show how one element acts on another. For now, let it suffice to know what the natural power of the air is: that it itself is light and it has this in common with fire. That is to say, it is light insofar as it is something which cannot be apprehended by the senses. Therefore, we will not see the air which is between us and the wall, because the quickness of the eyes matches the lightness of the air, and because of this similitude it does not offer any resistance; nor is it sensed, because nothing comes about from things which are completely similar. But sense does occur when, because of some hindrance, one is slightly less. For example, when a ray of the sun enters through the window, we see the atoms shining in the air. This is because sight conforms itself to the air which has been thinned out by the sun, and it then perceives the atoms shining. But in another part of the house where the air is not pure or thinned out, sight again conforms itself to the air which is slower, and it does not perceive the atoms between itself and the wall. For it is not possible for sight to sense something to which it has conformed. For example if anyone should go under water with his eyes open, he would not see the water. ...But if some grosser object were placed in the water, he would see it.

Air is also mobile, although its parts do not move so quickly as those of fire, and so air is less light and quick. It has these two powers in common with fire, but it has a third property natural and proper to itself, namely that it sustains life; that is, it is continuously inhaled and exhaled by breathing, and thus it vivifies, because nothing lives without inhaling and exhaling air. For this reason certain philosophers, Virgil and others, have said that air is Soul. This power belongs to air alone. Water cannot be inhaled and exhaled in such a way as to sustain life, but on the contrary it destroys life. If one were to be in certain parts of the world, such as the torrid zone, or on the peaks of mountains and other high places, he would not last long because the air there is thin and closely approaches the nature of fire. And if thin air cannot sustain life, how much less can fire, which is even lighter. However, the air at lower altitudes can be breathed because of its relative slowness. Because of the nourishing quality of air, the people are fat where the air is slower and lean where it is thin. These three things are the substantial powers of air. Some people say that there are others, *viz.* that it possesses levity and darkness by nature. But these things are the result of concommitant attributes, about which we shall speak when we discuss its colors.

Water is corpulent. A corpulent body properly speaking is one which offers resistance to the senses and can be sensed. It has this power in common with earth. Also, it is mobile, although not so much as the two higher elements. For this reason, it easily yields to sense, and when we look into water our sight goes right through it to the bottom unless turbulent mud or other earthy matter blocks it. It has this power in common with the higher elements.

But there is one power it has all to itself, namely moistness. Moistness is the most abundant matter in the composition of all things. Whence things are even said to have had their origin in moisture, and for this reason the Ocean is called the Father of Things. But now we must diligently consider this power of water, because in what follows it will be very helpful to us in pointing out the composition of things. Water is moist; that is, it makes things wet, and it has this quality as a result of its slowness, which makes it stick to anything touching the senses. Nevertheless, because of its mutability it easily yields that which it has, since it is quite liquid. For although it is not so immobile as earth, it is not so liquid as the higher elements.

CHAPTER 3

ROBERT GROSSETESTE AND
SCIENTIFIC METHOD

By the first quarter of the thirteenth century, virtually the entire scientific corpus of Aristotle had been translated into Latin several times, from both Greek and Arabic. Treatises on optics, catoptrics, geometry, astronomy, astrology, zoology, psychology and mechanics by both Greek and Muslim authors were added to these, and after 1231 the excellent commentaries of Averroes on Aristotle's works became available. Although most of this had been translated during the twelfth century, for several generations it remained known to only a few of the more inquisitive scholars. Around the year 1200 this material began making its way into the curricula of the leading educational institutions, which were just at this time achieving the status of universities. Paris and Oxford were the leaders, but by no means the only participants, in this. Once these works became a part of the regular curriculum, they were subjected to systematic analysis and study, and their influence became correspondingly more far-reaching.

The nature of medieval science was thereby changed in two important ways. First, the naive certainty of the twelfth century that the results of rational inquiry could never conflict with revealed truth gave way to a somewhat more cautious use of the new material as its implications became more fully understood: it could be very useful, but it could also be dangerous. Thus much of the daring, originality and confidence of the preceding century was lost. But this was compensated for by the possession and full comprehension of the most impressive body of scientific knowledge the world had known until modern times. This new knowledge was arranged in neat compartments, it was presented in elegant, rational and sophisticated fashion, and it contained an enormous amount of factual information about the natural world as well as highly developed methods of investigating that world.

Secondly, by far the most important single author among this new material was Aristotle, and it was his categories of thought and his world view which dominated the European mind for at least the next 300 years, although these were greatly modified in the process of being assimilated. Aristotle had analyzed each thing in the world as consisting of matter and form, neither of which could exist without the other. Alteration of any kind he accounted for as being a movement from potency to act. A thing which existed of itself he named a substance, and anything which existed as a concomitant attribute of a substance was called an accident. The various things of this world were classified by genus, and each genus, through specific differentia, was divided into species. Change from potency into act could be accounted for by four "causes," the material, the efficient, the formal, and the final. It was this general framework within which most scientific thought took place until the seventeenth century. Its precision and comprehensiveness were an enormous boon at first, but before the thirteenth century had run its course, it had already become inadequate to answer the kinds of questions men were asking of nature.

One of the principal means by which medieval scientists were able eventually to outstrip their ancient European and Muslim teachers was the development of a more adequate method of scientific inquiry, and the principal figure in bringing this about was Robert Grosseteste.

Grosseteste was born about 1168, began lecturing at Oxford shortly before 1200, and studied theology, possibly at Paris, between 1209 and 1215. He was then appointed chancellor of Oxford—very likely its first—but continued to teach and write. In 1229 he became the first lecturer to the newly established Franciscan convent at Oxford, and in 1235 was elected Bishop of Lincoln, England's largest diocese and the one in which Oxford lay. He died in 1253. When he was a young man, the twelfth-century translating activity was at its peak and these works were just beginning to invade the curriculum. During his teaching days they were received into the mainstream of European thought, and Grosseteste himself was a leader in introducing them at Oxford. He translated many works from the Greek and provided commentaries to several of Aristotle's more difficult works as well as short treatises on different aspects of the Philosopher's thought. He was also well versed in the Latin versions of Muslim and Jewish works, being an avid student of Muslim astronomy and astrology in his youth and one of the first Latins to use the works of Averroes after their arrival in 1231.

His achievements in scientific method result largely from his understanding of the creation story in Genesis. Although it sounds somewhat preposterous to twentieth-century students, this "Light Metaphysics," as it came to be called, is of such major importance that a summary of it here is necessary.

Grosseteste held that in the beginning God created the first corporeal form, light, which had the property of instantaneously multiplying itself infinitely in every direction; and simple matter, an unextended substance. The original point of light was joined to unextended matter (since matter and form never exist apart from each other) and in its expansion drew matter out into spatial dimensions. The resulting universe was a sphere, extremely rare at the outside but dense and opaque near the center. It was finite in size because of the limitations of matter and because a simple non-dimensional substance multiplied an infinite number of times would result in a finite quantity. The matter of the periphery was completely actualized and capable of no further change. When this first perfect body, containing only first matter and first form, had been created, it diffused its light back to the center, where it gathered together the mass existing below the first body, again rarefying the outermost parts and making the center more dense. The second sphere was thus formed, and by a similar process all thirteen spheres of our sensible world (the nine celestial spheres and those of the four elements). On the outside of our universe, matter is completely actualized and capable of no further change, while at the center actualization is less, and matter remains susceptible of taking on a great variety of forms. Every subsequent form, either substantial or accidental, is generated from the first form, and every privation is generated from the privation of light. Grosseteste considered light to be the efficient cause of local motion as well as the principle of intelligibility in the created universe.

Most of Grosseteste's other views either imply, or are derived from, his Light Metaphysics. In a series of scientific works written between 1220 and 1235 he progressively developed his characteristic method of investigating nature, and during this same period he elaborated this method from a theoretical standpoint in his commentaries on Aristotle's *Physics* and *Posterior Analytics* and in two short works, *On Lines, Angles and Figures* and *The Nature of Places*. According to this method, any complex phenomenon of nature must first be analyzed into its simplest components. Then the investigator, relying on his scientific intuition, must frame a hypothesis which would show how these elements are combined so that they actually produce the phenomenon under investigation. In addition to this framework, Grosseteste employed experiments (although these were often ones of which he had merely read or were simply appeals to ordinary experience) as an integral part of his investigation: as aids in accomplishing his analysis, as suggestions in framing his explanatory hypotheses, and as tests of the truth or falsity of a hypothesis. He also insisted that no accurate knowledge could be had of nature without mathematics, holding that since light is the cause of local motion, and light behaves according to geometric rules, therefore all local motion can be described mathematically. He also took over from Aristotle the principles of the uniformity and economy of nature, and he formulated, though he did not use, the principle that an experimental universal may be obtained from the observation that a given effect always results from a particular cause. When one controls his observations by eliminating any other possible cause of the effect, he may arrive at an experimental universal of provisional truth.

His scientific works cover a broad range of topics. While he was still quite young he wrote an astrological work, *On Prognostication*, but later denounced astrology as a fraud and a delusion of the devil. He wrote five works on astronomy. *On Supercelestial Motion* is an introduction to Aristotelian astronomy: *On the Sphere*, written before 1230, and perhaps antedating the better-known work of Sacrobosco, is an excellent brief introduction to Ptolemaic astronomy. The other three works are primarily concerned with reforming the Julian calendar, which was nearly four days in error in his time. Using the works of Ptolemy, Al-Battani and Ibn Thebit, he worked out a program for calendar reform which continued to find supporters until it was finally largely incorporated in the Gregorian reform of 1582. He also wrote works on *The Generation of the*

Stars, Sound, The Impressions of the Elements, The Tides, Comets, The Heat of the Sun, Color, and *The Rainbow.*

The Impression of the Elements is one of Grosseteste's earlier scientific works, having been written shortly after 1220, and is not nearly so good as his more advanced investigations. It suffers particularly from the fact that he had not yet acquired the knowledge of reflection and refraction, which give the later works so much of their value, and he had not yet developed his concept of the incorporation of solar rays in dense transparent media, of which he was to make such fruitful use in *The Heat of the Sun* and *The Tides.* However, the main features of his scientific method (aside from the use of mathematics) are clearly in evidence, and they strongly differentiate this treatise from similar works composed during the twelfth century.

Robert Grosseteste, *The Impressions of the Elements.* Translated from the text of Ludwig Baur, *Die philosophischen Werke des Robert Grosseteste, Bischofs von Lincoln* (Münster, 1912), pp. 87–89.

As James attests in his canonical letter: *Every good endowment and every perfect gift is from above, coming down from the Father of lights with whom there is no variation or shadow due to change.* [James, I: 17] But this must necessarily be done to some immediately, to others mediately. Therefore philosophers, although not understanding the matter perfectly, since they ought not to be ignorant of the world of nature, correctly say that the rays of heavenly bodies falling on corporeal things are the foremost cause of change in these bodies, because reflected and condensed rays are the cause of heat generated among us. Evidence of this is the fact that there is greater heat in valleys than on mountains. For this reason, snow remains on the mountains longer than in the valleys, and on the very highest mountains it remains perpetually. And note that it makes no difference that the sun in itself is hot. For if the solar body were hot in itself and its heat stirred up heat in lower things, then the closer a thing were to it the hotter it would be, and on mountain tops there would be greater heat than in valleys, and in the upper and middle layers of air there would be greater heat than in the lower. But we see that the opposite of this is true, because snow remains on mountain tops but not in valleys; and hail is formed in the upper layer of air and rain in the lower. Further evidence of the same thing is that birds of prey fly high in the summertime to cool themselves and eagles fly high to moderate the heat caused by their motion. And cranes and many other birds come down into the valleys to escape the cold, and when it gets too hot they return to the mountains. All these things are evidence of the fact that heat does not issue forth directly from the solar body, but is caused by a reflection and condensation of rays.

Now that we have established this, it is clear that the rays descend into the depths of water, since water is a transparent body like air, hornglass, and glass. Therefore, in the depths of the water there is a reflection, and so there is greater heat in the depths than on the surface. For this reason fish go to the bottom in winter but come to the surface in summer; and water freezes on the surface but not in the depths.

If anyone should ask why water congeals when it is made very cold, since coldness is its natural potency along with (apparently) humidity and fluidity, one can reply to this question that all water is naturally cold, but not fluid. In fact its own nature is rather to be congealed. Its fluidity results from enclosed heat.

Also, rays reflected from a concave mirror generate fire, and flax placed opposite the mirror is set on fire. Now that we have established that heat arises from the condensation of rays, it is clear that when they are condensed in the depths of water, the water is heated, and is heated so much that it cannot retain the nature of water. Therefore it passes over to the nature of air. But since it is not the nature of air to be lower than water, it rises in a bubble above the water. When many bubbles rise above the water at the same time, they maintain themselves because of their humid nature, and from these come vapor or steam from which clouds are made.

But since this generation of bubbles takes place in the depths of the water, some of the bubbles pass through channels of the earth, some remain in the water, and some rise above the water. We will first discuss those which rise. If anyone wishes actually to see this happen, he should put clear water in a clear vessel and place it over a flame. He will plainly see the bubbles generated and ascending because of the heat of the fire. The method of generation of the bubbles is the same in both cases.

We must also note that there are earth and fire along with the air and the bubble. Therefore, there are all four elements present in such a bubble: earthiness, because of the place of generation; the generated air; the nature of fire in the generation of heat; and water for obvious reasons. And so, this is much like the first generation of the elements and their first admixture.

When water predominates in the generated bubbles they are called "humid vapor;" when earth predominates they are called "dry smoke;" when air predominates they will be a dense vapor. Therefore, the bubbles are more subtle or more gross according to the subtlety or grossness of the generating heat, for if the heat were great and gross, it would generate large, gross and heavy bubbles, which seldom rise beyond the surface of the water, and there they are broken and the heat is dissipated. And the more subtle the heat, the more subtle the bubble and the weaker the heat. For this reason the bubbles do not separate from the surface of the earth, and they flutter upwards from it in the valleys. This occurs in the morning and the evening, when the heat is weak, and in this way a cloud is formed. And when these little bubbles lose their heat, they fall to the surface of the earth and become dew. But if the heat is greater, it makes these bubbles—or cloud—rise to the first of the three layers of air. Here the bubbles which make up the cloud lose their heat one by one, and they fall as drops of rain. Note that dew and rain differ both in respect of size and in places of their generation. But when a cloud rises to the second layer of air there is a greater loss of heat, and the bubbles lose their heat successively only, not suddenly, so that that which is soft is relinquished, just like wool, and becomes snow. But if the cloud is suddenly driven upward to the second layer of air, it suddenly loses its heat and becomes a round stone, or hail. This occurs especially when the heat is great. Frost however differs from a cloud just as rain differs from mist.

The Heat of the Sun has much in common with *The Impressions of The Elements*, but it was written about ten years later and shows how Grosseteste improved in the meantime. It is an excellent example of his scientific method actually employed in the solution of a problem.

Robert Grosseteste, *The Heat of the Sun*. Translated from the Latin text of Ludwig Baur, *Die philosophischen Werke des Robert Grosseteste, Bischofs von Lincoln* (Münster i. W., 1912), pp. 79–84.

Since we wish to know how the sun generates heat, let us first treat the more general question: How many ways can heat be generated? And although there are three things from which heat can be generated, namely a hot object, motion, and a concentration of rays, nevertheless we should understand that the heat in these things is of a single kind; and, by this single kind of heat, a single kind of effect is produced. ... Therefore, let us seek out this single kind of cause.

In all of these three cases, the proximate cause of heat is scattering. And so when a hot object generates heat, it does so by the scattering of bits of matter. But how scattering occurs in the cases of motion and a concentration of rays is difficult to see.

Local motion, from which heat is generated, is divided into natural motion and violent motion; and natural motion is divided into straight and circular motion. Let us speak first about violent motion, or about a heavy object moved violently. A heavy body can be moved violently in three different ways; either up, or down, or down but not directly toward the center of the earth. In all of these cases, it is clear that scattering occurs in a violently moved object as a result of motion. For in a violently moved object, there are two forces acting, namely natural and violent motion, which move each part of the moving object in different directions. This inclination of the parts to move in different directions results in scattering, and because of the scattering, such a violently moved object becomes hot. And because in the first way of moving violently (that is, straight up) there is the greatest opposition of the two moving forces—since they move in completely opposite directions—the greatest scattering and consequently the greatest heat result. But in the second and third ways, the heat is only moderate. This is utterly clear both to reason and experience.

This same thing is true in natural motion, for heat is generated in the motion of an object moving downward naturally. There are actually two forces, one natural, the other violent, acting on each part of such an object. It is obvious that there is a natural force involved here; but I shall prove that there is also a violent one: Any heavy object which moves downward, but not directly toward the center of the earth, is moved violently to some extent. All parts of a heavy object do not move directly toward the center of the earth. Therefore, all the parts of a heavy object are moved violently. I prove the minor premiss thus: The parts of a heavy body always maintain the same distance from each other in the whole body. Therefore, when they move downward by the motion of the whole body, they move along parallel lines, which are always the same distance apart. But parallel lines extended infinitely in either direction never meet. Therefore, the parts of a naturally moved heavy body move downwards along lines which do not meet. Therefore, they do not move directly toward the center of the earth, for if they did, they would be moving along lines which converge at the center of the earth. Thus the principle has been made clear that there are two forces acting on each part of a body moving downward naturally, which incline it toward different directions. But the opposition between such inclinations is much less than that between the parts of a body which is moved violently. And therefore, among all the things that generate heat, natural motion generates the least natural heat. By these arguments, it is clear that heat is generated by natural motion in a straight line and by violent motion, and that this is done by the same kind of cause as when heat is generated by a hot object.

Similar arguments can prove that this is also true of the third way of generating heat. That some heat is generated by the single kind of cause of heat by means of a concentration of rays is made clear in Euclid's *Catoptrica*, where it is stated that a combustible substance can be set on fire by a concave mirror aimed toward the sun. This happens because of scattering. For a ray is more incorporated in a dense medium than in a rare one — and we do not mean here total incorporation as in the case of heat, but a certain partial incorporation. And because of this incorporation, the ray carries along parts of the air with it. Then when the rays are all concentrated at one point, each ray having come there along its own straightline path, there will be a very great separation of the air in different directions around that point; and thus there will be scattering and consequently heat. Thus, therefore, it is clear that heat is present in each of these three kinds of generation because of a single kind of cause.

If then the sun generates heat, it will do so as a hot object does, as motion does, or as a concentration of rays does. That it does not generate heat as a hot object does is clear thus: Aristotle proves in Book VII of the *Physics* that a thing which causes change and the thing which is changed by it must be in immediate contact. Therefore, if there were some medium between the agent which originates the change and that which is ultimately changed, that medium would first have to be heated by the hot sun before it could cause heat in the ultimate subject. Otherwise, the agent and recipient of change would not be in immediate contact. Therefore, since there are several media between the sun and the air, and the fifth element is nearest the sun, it follows that the fifth element would have to be heated up by the hot sun before the air could be so heated. But this is impossible, because if the fifth element could be changed it would be corruptible. Therefore, the first way is impossible, namely that the sun causes heat in the manner of a hot object.

That heat is not generated by the motion of the sun is also clear. Motion only generates heat when each separate part of the moved object tends to move in different directions. But when something moves circularly (and not violently), each part has the same inclination as the whole, and there is no tension. The tendency of each part is to move circularly. Therefore no heat is generated by circular motion....

The only theory remaining is that the sun generates heat through a concentration of rays. This is clear thus: The rays of the sun in the transparent body of the air are incorporated in it to some extent because of the nature of a dense body. But the rays of the sun falling upon the surface of the earth (whether on a flat, convex, or concave portion of it) are reflected at equal angles, as is shown in the last part of Euclid's *Catoptrica*. Therefore, if the rays fall perpendicularly, they are reflected perpendicularly. And thus the incident and reflected rays go along the same path but in completely opposite directions, and then there is a maximum of scattering. This occurs on the equator when the sun passes over the zenith of this region; in places lying between the tropics of Cancer and Capricorn the sun's rays must fall perpendicularly twice a year; but in a place lying on the tropics of Cancer or Capricorn the sun is at the zenith only once a year, and so its rays fall perpendicularly only once. When this happens, there is a maximum of scattering of the particles of air, and the greatest heat. And this is a violent scattering such as occurs from a concentration of rays refracted by a spherical body or reflected from a concave mirror, except that in these examples the ray is not reflected in a completely opposite direction, as is the case with a perpendicular ray of the sun.

But in regions which lie beyond the tropic of Cancer, since the sun does not come far enough north ever to be directly overhead, the sun's rays fall at less than a right angle and are reflected at the same angle and therefore not in completely opposite directions. And the farther a place is from the equator, the more obtuse will be the angle at which the sun's rays fall and are reflected, and the less nearly opposite will be the paths of the incident and reflected rays, and the less scattering will occur, and the less heat is generated. And this is verified by experience.

But if it should be asked why the sun's rays do not generate heat in the fifth element, two answers can be given. First, the rays are not reflected in such a way that they intersect there. And secondly, even if they did intersect by being reflected in completely opposite directions, heat would not be generated. For since there is nothing of the nature of a dense body in the fifth element, the solar rays are not incorporated in it in any way, and so they are not able to scatter its parts. For this reason also, in the highest layer of air, where the air is thinnest, the least heat is generated; this is made clear by observation. For on the tops of mountains, where the solar rays are brighter than in the valleys, there is a great deal of snow, even though the reflection of rays occurs there just as it does in the valley. But because of the thinness of the air there the density of the air is slight, and there is a correspondingly slight incorporation of the light with the air, and therefore a slight scattering of the parts of air as a result of the condensation of rays. But in the valley, there is a greater incorporation of rays and therefore greater scattering and heat.

CHAPTER 4

THE TIDES

The tides had been the object of scientific investigation at least since the time of Aristotle, and a smattering of the knowledge of the great Hellenistic investigators had survived in the Latin handbook tradition. In addition, Bede, in his *De temporum ratione*, had made some very important observations on tidal phenomena, as we have mentioned before, including their relationship to the motion of the moon and the principle of the establishment of port. During the twelfth century, the Spanish Moslem Al Bitruji (Alpetragius), as a consequence of his "Aristotelian" astronomy, proposed a new explanation of the tides (summarized and refuted in the selection which follows) which Michael Scot translated into Latin about 1217. Gerald of Wales, an acquaintance of Grosseteste, compiled some important data on tidal phenomena on the opposite coasts of England and Ireland, and although he noted their relationship to the motion of the moon, resorted to the Maelstrom theory to explain them. During this century too the works of Ptolemy and other Greek and Arabic writers had been translated into Latin. By the early thirteenth century then, there was a long tradition of tidal investigation accessible to scientists, a body of reliable data and several proposed explanations. As part of his teaching activity at Oxford, Robert Grosseteste composed a remarkable short study of the tides. Fully cognizant of the tradition of tidal studies, Grosseteste, instead of choosing among already existing alternatives, undertook an original investigation using new methods and techniques.

According to the modern theory, the moon is the principal factor in causing the tides. The "gravitational pull" of the moon causes the earth — and especially the seas, since they are fluid — to bulge out toward the moon (because the gravitational pull is strongest in that direction) and away from the moon on the other side of the earth (since the gravitational pull is weakest in that direction). Therefore, there will be two high tides and two low tides each time the moon makes a complete circuit of the earth (approximately 24 hours, 50 minutes). The earth's rotation causes a delay of up to two hours between the moon's meridian passage and the high tide; the exact time interval varies for different points on the earth's surface. The tide will also be affected by the moon's distance from the earth (it will be higher when the moon is closer) and by its position relative to the sun. Since the sun also exerts a considerable gravitational pull on the earth, when it and the moon are in opposition or conjunction (that is, when they lie approximately on the same straight line either on opposite sides or on the same side of the earth), the tide will be increased; when they are at right angles to each other, the height of the tide will be decreased.

Grosseteste's explanation accounts quite well for the observed phenomena. But since there was no theory of universal gravitation available to a thirteenth-century scientist, he had to search elsewhere for an explanation of the mechanics of the tides. Consistently with his Metaphysics of Light, he assumed that it was the quantity of light in the moon and the angle at which its rays fell on the sea that accounted for the variations in the height of the tides, and that the tidal bulge was caused by the moon's light freeing gaseous matter which had been trapped under the water.

Selections from Robert Grosseteste, *An Inquiry into the Causes of the Tides.* Reprinted by permission from R. C. Dales, "The Text of Robert Grosseteste's *Questio de fluxu et refluxu maris* with an English Translation," *Isis*, LVII (1966), 455–474.

In our investigation of the tides, let us first determine the material cause, which is twofold, namely general and specific.

The general cause is this: The spheres of the four elements are so arranged that earth is in the center and fire on the periphery, while water and air occupy places between these two. These elements are of such a nature that they can be moved by rarefaction and condensation. But water and air share this capacity to a greater extent than fire and earth; therefore they are better suited to be moved. The minor premiss is proved thus: Condensation is the motion of the parts of matter toward the center of the universe, and rarefaction is the motion of the parts of matter toward its periphery. Therefore, that which is in the very center, such as earth, is not further condensable; and that which is at the periphery, such as fire, cannot be further rarefied. However, those things which are between the center and the periphery, such as water and air, can be rarefied and condensed.

Now for the specific material cause: Every motion of a generable or corruptible thing takes place in water and air. Let us omit the motions of the air and speak about the motions of water. This investigation has three parts. The first part concerns the material and efficient cause of such motions. The second part concerns the reasons for the increase and decrease of the tides. The third part concerns the three kinds of seas: those which have tides, those which do not have tides, and those which have very small tides but do not seem to have any.

Turning our attention to the first part, let us first determine the efficient cause. This efficient cause must necessarily be a power of the sky or the power of a star in the sky. For an element is not moved by itself or by another element, but we ought to reduce every motion of the elements to an efficient cause which is an immaterial species.

In this connection, Alpetragius says that all the lower spheres, as far as the sphere of water, are moved from east to west by the power of the outermost sky. But the lower any particular sphere is, the less power it will receive from the outermost sky, because this power is the power of a body and diminishes with distance. The earth, however, both because it is farthest away from the outermost sphere and because of its heaviness, remains completely immobile. Therefore, the water of the sea is moved by the power of the outermost sphere from east to west, and from this, Alpetragius says, a crashing together of the waters occurs and consequently a high tide. But the water returns to its original place because of its heaviness, and when it has all returned it begins to be moved again and to rise as far as its heaviness will permit, and then again it returns. And these two high and low tides take place in a time greater than one day and night.

That this is false is clear thus: In the outermost sky, the parts which are moved most rapidly are the parts existing on the equinoctial circle, directly under which, on the surface of the earth, is the great sea called Ocean, surrounding the land; and this is the source of all seas. Therefore, if in one part of the sky there were a motive force and in another part there were not, when this part of the sky rose one part of the sea would be moved by it and another would not. If, therefore, one part of the sea should be moved and another not, the one part would resist the other, and one part would have to be elevated above the other and occupy a greater place. But since the sea is circular, and the motive power operates equally in every part of the equinoctial circle, therefore all the parts of the sea under each single part of the equinoctial circle will be moved equally in the same direction. Therefore, one part of the sea will not resist the motion of the other part, and there will be no reason why the water should occupy a greater place. Likewise, if the sea were moved in this manner, only local motion would result from such a cause. But we perceive by experience that the rise and fall of the sea results from its condensation and rarefaction. For ships in the sea draw less water at high tide than at low tide, and this is because the low tide is caused by rarefaction and the high tide by condensation. Also, the water is found to be hotter at high tide that at low, and this can only be because of the greater subtlety of its parts. These arguments and experiments are sufficient to prove Alpetragius' explanation false.

That the moon alone, and not some planet or image of the fixed stars, is the cause of this motion is clear for this reason, that the motion of the sea follows the motion of the moon more closely than that of any other planet; and there is a definite proportion between these two motions, as will be made clear below. The arguments of the astronomers lead us to a similar conclusion. They point out that there are two great luminous bodies, the sun and the moon. Of these two, the sun is primarily responsible for motions which take place in the air, and the moon for those which take place in water, since in these two spheres (i.e., air and water) the generation and corruption of all living things takes place, and these two luminous bodies are the principal causes of every generation and corruption. Also we know by experience that, of all the heavenly substances, the moon exercises the greatest control over moist and cold bodies. Thus certain people are called lunatics because, when the moon wanes, they suffer a diminution of the cerebrum, since the cerebrum is a cold and moist substance. These arguments are enough to show that the moon is the sufficient and efficient cause of the tides.

But how it causes them, we must still discover. When the moon rises on the horizon of any sea, it first casts its luminous rays on the center of that sea and, strongly impressing its power, it moves this sea, and this motion increases until the moon arrives at the meridian. But when it passes over the meridian its effective power is diminished, and the sea recedes toward its original place until the moon has set. When the moon again passes over from the west to the middle of the sky under the earth, the sea is increased; and when the moon passes from the middle of the sky under the earth, the sea is decreased until the moon again begins to rise. And thus in one revolution of the moon from rise to rise, two high tides occur in that place over whose horizon the moon has risen.

However, the explanation of why the high and low tides correspond to these four quarters of the sky is very difficult. It is clear enough from what has been said why the high tide occurs between the moon's rise and its arrival at the meridian and why the low tide occurs when the moon is passing from the meridian to the western horizon. For while it is rising it impresses its power more strongly than when it is setting. But the other high and low tide take place when the moon is in the two quarters below the horizon, and its light cannot act on the sea. Since heavenly bodies can only act on lower bodies by their light, it is doubtful how the moon can be the cause of the motion of the sea. The astronomers answer this by saying that opposite quarters of the sky have similar effects, but whether this is true remains to be proved and is in need of further investigation.

Therefore the time taken by two complete high and low tides exceeds the time of one day and night by the same amount as does one revolution of the moon from rise to rise. You must find out, therefore, by how many hours the rise of the moon follows or precedes the rise of the sun. We know by how many hours the beginning of the rise of the sea follows or precedes the beginning of the day and that one lunation contains twenty-nine days and a few minutes. Therefore, in seven and a quarter days and a few minutes the moon will be one quarter of a circle east of the sun; whence in one-third of this time, when the sun is beginning to rise, the moon will be in the middle of the sky under the earth, and then the tide will begin to recede, since at the beginning of this time the tide began to rise. In twice this time, however, when the sun is in the east and the moon is in the west, the tide will begin to rise. But if the moon should be more than three quarters away from the sun, when the sun begins to rise, the moon will be in the middle of the sky above the earth, and the tide will begin to recede. However, when the moon is again in conjunction with the sun, the beginning of the day and the beginning of the rise will occur at the same time. And this is what sailors say, that if the beginning of the rise coincides with the beginning of the day, seven days later the beginning of the recession will coincide with the beginning of the day. But this rise and fall in the same times does not appear on the shores of seas which are far away from the middle of that sea where the rise and fall of the tides originate. And you should know that on the shores nearer to the south and east, the rise takes place more quickly. Let this suffice concerning the efficient cause.

Now it remains to discuss the specific material cause, and it is this: Waters tend to come together in a deep and broad place, and in these waters is much matter of vapors and winds. Whence the moon, rising and impressing its power, generates many vapors in these waters and stirs up the winds. But this water, because of the heaviness and viscosity of its parts, will not let them escape, but rumbles and becomes swollen because of the trapped vapors. But since fresh water is subtle and its parts are penetrable, when any vapor is generated in it, it is expelled at once; hence it does not, strictly speaking, have tides.

Also, if a heavenly body acts on lower bodies only by its light rays, since these light rays are in some manner incorporated with the elements, they intersect themselves at one point when they are reflected, and thus they generate heat by scattering the parts of matter in various directions, as is clear from the last proposition of Euclid's *Catoptrica*. Therefore the subtler the matter, the less heat will be generated by the ray. For this reason, snow remains longer on the tops of mountains than in valleys because the air is subtler there. Therefore if fresh water is much more subtle than sea water, the lunar rays will be much less incorporated in it than in sea water, and thus they will generate less heat and a smaller effect. Let these remarks suffice concerning the first part of our investigation.

II

Let us go on to the second part, in which we are to seek the reason why sometimes the tide increases and becomes strong and at other times it decreases and becomes weak. There are eight reasons for this.

The first is that when the sun and moon are in conjunction, the power of the moon becomes stronger, and the tide increases and becomes strong. But as the moon recedes from the sun, the tide decreases until the moon is one quarter of a circle away from the sun. From this point until the moon is a half-circle away from the sun, its power increases, not because of the moon's nearness to the sun but because of the increase of light in it. And again, from this point until the moon is three quarters away from the sun, the tide decreases; and then it increases again until the sun and moon are in conjunction. And this occurs proportionately according to the four parts of one month corresponding to the four parts of one day.

The second reason is the comparison of the diverse motion of the moon with the mean motion of the moon. For if the motion of the sea is caused by the motion of the moon, then the mean motion of the sea should correspond to the mean motion of the moon, and the diverse motion of the sea to the diverse motion of the moon. And so, when the moon's true motion varies from the mean motion in one direction, the tide is increased; and when it varies in the other direction, the tide is lessened.

The third reason is the approach of the moon to the *aux* of its circle, that is to its longitude farthest from earth, for then its power on the sea is diminished because of its greater distance from earth. And when it approaches the point opposite the *aux*, that is its longitude nearest the earth, its power increases, and then the rise of the sea is strong.

The fourth reason is the southward declination of the moon in latitude from the zodiacal circle, for then it approaches the middle of the Ocean in which the strong rise begins. However, when the moon declines north of the zodiac, the high tide is lower because of the moon's greater distance from the middle of the Ocean.

The fifth reason is the existence of the moon in the southern or northern signs of the zodiac, and this is a particular cause of the high tide. For when the moon is in the southern signs, it increases the high tide in the corresponding place on earth. But when it is in the northern signs, it increases the tide in the northern sea.

The sixth reason is the days which are called "Egyptian Days" by the ancients, because the Egyptians first discovered them. But let us omit this reason for the present because many effects lie hidden in these days.

The seventh reason is the help the sun gives the moon. From the spring equinox to the summer solstice, the sun increases the tide. But from that point to the autumnal equinox, it diminishes it. From the autumnal equinox, however, until the winter solstice, it increases it again. And from there to the spring equinox, it diminishes it. And there are these four quarters of the year, just as earlier we noted the four quarters of the month corresponding to the four quarters of the day.

The eighth reason is the wind. When it flows in the direction of the tide, it increases it; but when it blows in the opposite direction, it decreases it. ...

CHAPTER 5

STUDIES OF THE RAINBOW

One of the most impressive results of the employment of the mathematical experimental method in the Middle Ages was the series of solutions proposed for the very difficult problem of the rainbow. The starting point of these studies was Aristotle's *Meteorology*, and the turning point was the work of Robert Grosseteste. They culminated in the work of Theodoric of Freiberg shortly after the beginning of the fourteenth century.

One of the most striking aspects of the study of the rainbow in the thirteenth century is the originality and imagination which accompanied the experimental work. It was generally agreed that even the great philosopher, Aristotle, had not given an adequate explanation, and that his commentators, both Greek and Arabic, of Antiquity and the Middle Ages, had also fallen short. The progress of the study of the rainbow from Grosseteste to Theodoric resulted from consecutive criticisms of what had been done before and from an accumulation of insights and experimental data, as well as from the application of an experimental method ultimately derived from Grosseteste, but greatly improved as the century went on.

In his *Meteorology* [Book III, 1–5. See also Carl Boyer, *The Rainbow: From Myth to Mathematics* (New York and London, 1959), pp. 39–54 and A. Sayili, "The Aristotelian Explanation of the Rainbow," *Isis*, XXX (1939), 65–83] Aristotle had worked out an ingenious explanation of the rainbow, which dominated all treatments of the subject in Antiquity. He asserted that the rainbow was caused by the sun's rays being reflected from a cloud to the eye of the observer. (Actually he used the vocabulary of the extramission theory, whereby the sight is reflected from the cloud to the sun, but the geometry of the explanation is the same for either case). The reason that the bow was a band of colors and not simply an image of the sun was because the cloud was considered to be made up of tiny drops of water and air, each too small to reflect the sun itself, but capable only of reflecting its colors. These drops were considered, though, to constitute a single reflecting surface, much like a rough wall, and Aristotle never felt obliged to investigate the behavior of light falling upon a single drop. To account for the shape of the bow, he resorted to an ingenious mathematical device which could "save the appearances" but could hardly have been intended by him to be in accordance with physical reality. He posited a hemisphere, called the "meteorological hemisphere" by Cårl Boyer, with its base on the plane of the horizon with the observer situated at the center and the sun at some point on the circumference. Assuming the sun to be on the horizon (at which time the bow would have the maximum elevation), the rays would be reflected from a cloud also situated on the circumference of the meteorological hemisphere (and consequently at the same distance from the observer as the sun!) to the eye of the observer. Assuming further that the lengths of the sides of the triangle thus formed are a constant, if one rotated the triangle around its base on the plane of the horizon, an arc of a circle would be described on the meteorological hemisphere. [See figure 1.] This accounts for the shape and position of the bow.

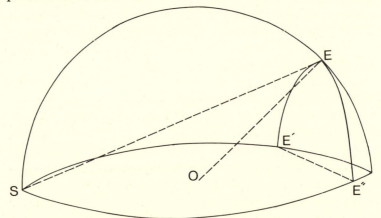

Fig. 1.   Simplified diagram of Aristotle's theory of the rainbow.

In order to explain why only this narrow bow-shaped band of colors appears, Aristotle asserted that only those rays which were reflected at the proper distance (depending on the constant ratio between the lines from the sun to the cloud and from the cloud to the observer) would produce the rainbow colors, and he gives precise directions on how to determine this distance. The colors formed result from the weakening of the sun's rays, an effect which can be accomplished by reflection or refraction, by traversing a long distance, or by the weakness of sight. The brightest color after white was red, and so the red band was caused by the rays which traversed the shortest distance (and so were least weakened), the blue traversed the longest distance, and green was in between. These were the only true colors in the rainbow, Aristotle claimed, although yellow may appear from time to time because of contrast.

Aristotle also noted the existence of the secondary rainbow, but his remarks on it are too obscure to give us any clear notion of what he may have thought about it.

This explanation was the starting point for all subsequent treatments of the rainbow. In Antiquity not much was added to what Aristotle had said. Seneca, in his *Natural Questions*, largely based his account on Aristotle; he emphasized the importance of the individual drops of moisture and suggested, in passing, the analogy between the sun's rays causing the rainbow by passing through a cloud and the behavior of light passing through a glass sphere filled with water.

During the twelfth century, Aristotle's *Meteorology* was translated into Latin by several authors, and the English scientist, Alfred of Sareshel, provided it with a much-used commentary. In about 1235, Robert Grosseteste attempted a daring solution of the problem of a rainbow, based on the assumption that it was caused by refracted rather than reflected light. Crucial aspects of Grosseteste's theory of the rainbow are so ambiguously worded that there is considerable disagreement among scholars as to just what he was proposing [See, for example, David C. Lindberg, "Roger Bacon's Theory of the Rainbow: Progress or Regress?" *Isis*, LVII (1966), 235–248; Bruce S. Eastwood, "Robert Grosseteste's Theory of the Rainbow," *Archives internationales d'histoire des sciences*, XIX (1966), 313–332; and R. C. Dales, "Robert Grosseteste's Scientific Works," *Isis*, LII (1961), 398–399], but the principal mechanisms are quite clear.

After noting that the science of the rainbow is the province of both physics, which is concerned with the experienced phenomenon itself, and optics, which is concerned with the reasons for the phenomenon, he considers the three methods of transition of a visual ray to the seen object: in a straight line, by reflection, or by refraction. He shows that the rainbow could not be caused by either of the first two and so must necessarily be caused by refracted light. He then proceeds to construct a mechanism which will explain how the sun's rays can be refracted in such a way as to produce a rainbow.

The rainbow, he says, is produced by the compound refraction of the sun's rays in the mist of a convex cloud. He then posits a cone of moisture descending from the cloud to the earth, rare at the top and more dense near the earth (but it is not clear in which direction the vertex of the cone lies). There will consequently be four transparent media through which the sun's rays must pass —the air containing the cloud, the cloud itself, the upper and rarer moisture, and the lower and denser moisture—and the light will be refracted as it passes from one of these to the other. The rays are refracted at the point of contact of the air and cloud, then of the cloud and rarer moisture. Because of these refractions, the rays run together in the denser part of the moisture and are refracted there again and spread out into a figure like the curved surface of a cone expanded in the direction opposite the sun. When the upper portion of the hollow cone of rays (at least half of which will always fall on the earth and consequently be invisible), falls on a second cloud, which is opposite the sun and acts as a screen, it produces a rainbow. [See figure 2]. This accounts for the bow's shape and position. The colors result from three factors: the density of the medium, the amount of light, and the clarity of the light. As the rays are progressively weakened by the series of refractions, the order of colors from red to violet appears. The role of refraction in this explanation is twofold: to cause the sun's rays to assume the shape of a cone; and, as a result of this, to weaken the light in such a way that the rainbow colors appear.

There are many obvious flaws in this explanation, but it was a work of the utmost importance. The problem of the rainbow is extremely difficult; its full solution did not come until the seven-

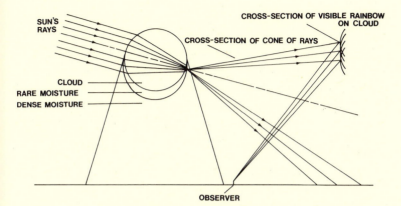

Fig. 2. One possible interpretation of Grosseteste's theory of the rainbow. From Bruce S. Eastwood, "Robert Grosseteste's Theory of the
Rainbow, "*Archives internationales d'histoire
des sciences,* XIX (1966), 320.

teenth century. It was Grosseteste who first emphasized the role
of refraction in the phenomenon, even though he did not see
precisely what that role was. His explanation has a high degree of
internal consistency, and its attempt to explain a complex phenomenon as the result of several simple ones is most impressive.
But most important is the fact that Grosseteste had made a radical
break from the Aristotelian tradition and had posed the problem
in such a way that the correct (or nearly correct) solution could
come about from a series of criticisms of his shortcomings.

Albert the Great, a somewhat younger contemporary of Grosseteste, wrote a lengthy and diffuse work on the rainbow, which,
while it did not present a carefully worked out theory, still aided
in the solution of the problem. He adopted Grosseteste's theory,
with some modifications. Taking his cue from Seneca, he emphasized the part played by the individual drops in the formation
of the rainbow, but he was not able to develop this suggestion. He
was an indefatigable if unsystematic experimenter. He very carefully describes what happens when one passes sunlight through
prisms, spherical balls, and a glass hemisphere filled with black

ink, and he reports having seen as many as four rainbows at once, although Aristotle had said there could be no more than two.

The first really significant criticism of Grosseteste's theory was offered by Roger Bacon in his *Opus maius*. Grosseteste had considered his explanation finished when he traced the sun's rays to the cloud "opposite the sun" acting as a screen. He seemed to feel that it was objectively situated there. Bacon pointed out that, to the contrary, the rainbow moves as the observer moves, and each observer sees a different rainbow; at all times (as Aristotle had noted) the sun, the observer, and the center of the bow are all situated on the same straight line. This, Bacon claims, is inconsistent with the laws of refraction, and as evidence of its inconsistency he calls attention to cases of images caused by refracted light—clouds, crystalline stones, a stick placed in water—which are stationary and do not move with the observer. Bacon also noted that a rainbow can occur in an irregular spray, such as that caused by splashing oars, where Grosseteste's four media and three refractions could not possibly exist. He pointed out that the refractions posited by Grosseteste would produce a solid cone of bright rays, and not just the surface of a cone. Following Aristotle, Seneca, and Albert, he emphasized the importance of the individual drops in causing the rainbow, but he was not able to go beyond Aristotle in describing that role. He did, however, realize that the bow was seen in a different set of drops by each observer. In explaining why the colors appeared in the particular drops they did and not in others, he followed Aristotle in asserting that this would only occur when there was the proper amount of light and the necessary darkness of the background. He added to this the observation that the colors appeared when light was reflected to the eye at the correct angle, but he could not explain why a given angle was "correct." Still, his insistence on this aspect of the phenomenon was of crucial importance to the solution of the problem. Bacon also provided the first bit of accurate measurement of any aspect of the rainbow: he gave the radius of the bow as 42°, a value very close to the modern one. And in a rare moment of modesty, he admitted that "many experiments are needed to determine the nature of the rainbow regarding both its color and its shape." In his criticism of the weaknesses of Grosseteste's explanation, his refocussing of attention on the individual drops of moisture, his body of experimental work, his realization that each observer saw a different bow, and his measurement of the bow's radius, Bacon significantly advanced the study of the rainbow, but his own theory was still sadly inadequate.

The next steps of basic importance were taken by the Polish scholar, Witelo, who was born about 1230 and wrote his *Optics* between 1270 and 1278. He had the advantage of knowing the *Treasury of Optics* of the great Muslim scholar, Alhazen, which had been translated into Latin around 1250. He was also undoubtedly familiar with the works of both Grosseteste and Bacon. He resumed Grosseteste's emphasis on refraction, but instead of the cumbersome cloud/cone of moisture/reflecting cloud mechanism, he concentrated on the behavior of light in individual drops. He investigated experimentally the paths of light through glass spheres filled with water (which he treated much like a microcosmic analogue to Grosseteste's cloud and mist), and he studied the production of colors by light rays refracted through hexagonal crystals.

Striving to construct a satisfactory theory, he seems to have borrowed suggestions from both Grosseteste and Bacon and then to have gone beyond either of them. He asserted that the rainbow was caused by both reflected and refracted light. Since clouds, he said, are composed of both moist and dry vapors, and light does not penetrate dry vapors but does pass through moist ones, some light would be reflected from the outer surface of the cloud (from drops of "dry" vapor) and some would penetrate to the interior of the cloud where they would be condensed by refraction and then reflected from other drops to the eye of the observer. The weakening of the sun's light, caused by both reflection and refraction, produced the colors of the bow, and this would happen only when the rays were reflected at the "correct" angle (42° more or less, depending on the density of the cloud).

From Ptolemy *via* Alhazen, Witelo learned how to make a device for measuring angles of refraction, and he repeated the directions for making this instrument in his *Optics*. From the same source, he knew Ptolemy's table of refractions from air to water or glass. Witelo greatly expanded this table, sometimes by actually measuring the angle of refraction, sometimes by assuming that light going from glass to air would follow the same path but in the opposite direction as light going from air to glass, and sometimes by extrapolation. For example, he gives angles of refraction from water to air for light incident at 50° to 80°, when in fact all the light would be reflected.

The high point of medieval rainbow studies came shortly after the turn of the fourteenth century in the work of the German Dominican, Theodoric of Freiberg. Theodoric was a philosopher and theologian of some note and had studied for several years at Paris, the intellectual center of Europe. In about 1304, at the suggestion of the Master-General of the Dominican order, he began writing down the results of his thoughts and experiments on the rainbow. [For a much fuller and more technical account of Theodoric's theory than what follows, see Carl Boyer, *The Rainbow*, pp. 110–124 and A. C. Crombie, *Robert Grosseteste*, pp. 233–259.]

The result was a lengthy, systematic treatise entitled *The Rainbow and Other Effects Caused by the Sun's Rays (De iride et radialibus impressionibus)*. It begins with a general treatment of the subject, enumerating the phenomena to be considered and classifying them according to the manner in which they are produced: by a single reflection, a single refraction, two refractions and one internal reflection, two refractions and two internal reflections, or total reflection from the common surface of the two media. His remarks on the roles of physics and optics in the science of the rainbow and his account of the three ways in which light can be transmitted (straight line, reflection, or refraction) strongly suggest that he knew Grosseteste's work on the rainbow. The second and most important part of the work deals with the primary rainbow, the third with the secondary rainbow, and the fourth with other phenomena caused by the reflection and/or refraction of the sun's rays.

Even though Theodoric's work is not without mistakes, it is a beautiful example of the employment of the mathematical experimental method which had been worked out at Oxford during the preceding century. He employed the principles of falsification and economy with great skill and to good effect. Although he mentions no Latin authors but Albert the Great, it is generally agreed that he was familiar with the major Latin works on the rainbow, as well as the Latin version of Alhazen's *Treasury of Optics*. His work went beyond his predecessors on two counts. First, it was based on a much fuller and more precise experimental foundation. And second, Theodoric had that flash of insight into the problem which no amount of experimentation could compensate for—he realized that the colors of the rainbow were formed by the refraction, internal reflection, and second refraction of rays of light in individual raindrops, [See figure 3] and that large numbers of drops close together, each reflecting one color to the eye (depending on the angle made by the sun's rays and the line of sight) constituted the rainbow [See figure 4]. He did a vast amount of experimental work using prisms and screens, tracing the path of refracted light, showing that colors can intersect inside a refracting medium and not lose their integrity, and investigating closely a large number of ways in which the rainbow colors can be produced either in nature or by contrived experiments.

Like Witelo, he experimented by passing sunlight through a glass flask filled with water. But unlike Witelo (who seemed to consider his flask as a small cloud), Theodoric clearly considered it to be a model of a single raindrop; and by covering up certain portions of the sphere one after the other he was able to trace the path of a ray after it entered the "drop," as well as the formation of colors during the process. He noted that the ray was refracted upon entering the drop—this is what would have been expected on the basis of traditional knowledge of refraction. But then he made the crucial discovery: that some of the light was reflected from the inner back surface of the drop and then refracted again as it came out the front surface into the air. It then occurred to him to hold the flask up in front of him and raise and lower it slightly. He noticed that different colors appeared depending on the elevation of the flask. This was the most important single discovery in the whole long history of the study of the rainbow.

It did not however constitute the entire solution. Theodoric realized correctly that all the drops reflecting the sun's rays at the

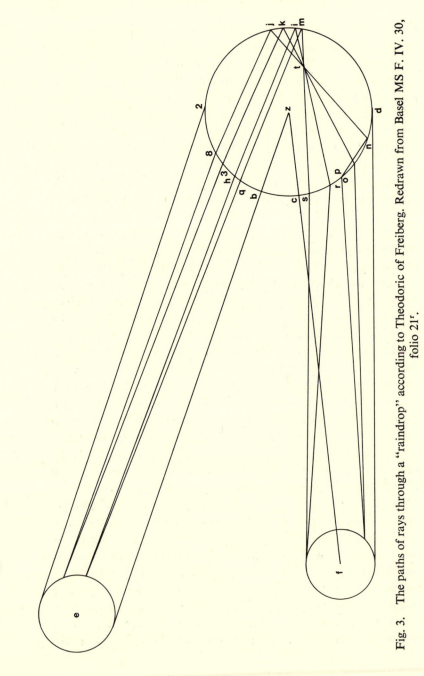

Fig. 3. The paths of rays through a "raindrop" according to Theodoric of Freiberg. Redrawn from Basel MS F. IV. 30, folio 21ʳ.

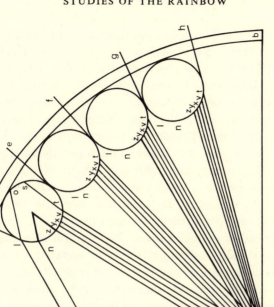

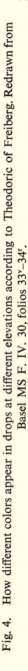

Fig. 4. How different colors appear in drops at different elevations according to Theodoric of Freiberg. Redrawn from Basel MS F. IV. 30, folios 33ᵛ–34ʳ.

same angle to the eye would appear to have the same color and would have the shape of a bow, but he incorrectly followed Aristotle in thinking that this depended on the "effective ratio" between the lines from the sun to the drops and from the eye to the drops. He also erred in concentrating his attention on the size of the arc between the incident and emergent rays. There are two more ways in which Theodoric's theory was faulty. First, his theory of colors was still based (with slight modifications) on the Aristotelian notion that color is produced by the weakening of white light. He found it necessary to make a number of *ad hoc* adjustments in the theory to make it accord with experimental evidence. Secondly, he inexplicably used the value 22° rather than 42° for the radius of the rainbow. No one will ever know why, but the most likely suggestion is that he copied a wrong figure early in his computations and never went back to check it.

Having correctly explained the mechanism by which the primary bow is formed, Theodoric proceeded to give the reasons for the secondary bow: light entering near the bottom of the drop is reflected twice inside the drop, thus being rotated counter-clockwise, before it emerges from the top of the drop, from where it goes to the eye of the observer. In short, it is formed just as the primary bow is, but with one more internal reflection [See figures 5 and 6]. He was hard put however to give a convincing

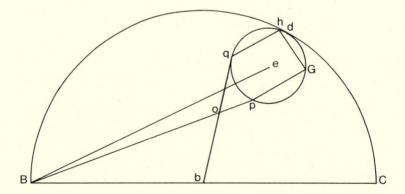

Figs. 5 and 6.   Theodoric's illustrations of how the secondary rainbow is produced. Redrawn from Basel MS F. IV. 30, folios 38ʳ and 40ʳ.

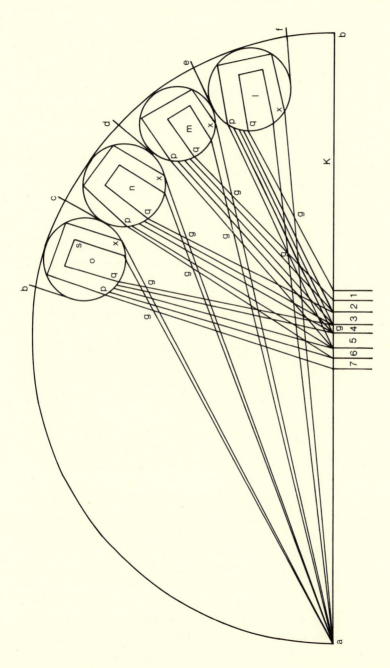

reason for the reversal of the order of colors in the secondary bow, or to explain the "Aphrodisian paradox," that is, the effect, noted by Alexander of Aphrodidias, a second century A.D. commentator on Aristotle, that the area of the sky between the red upper band of the primary rainbow and the red lower band of the secondary bow was dark, whereas one would expect it to be bright, since red is the "brightest" (i.e., least weakened) color.

Selections reprinted by permission from Carl B. Boyer, "The Theory of the Rainbow: Medieval Triumph and Failure," *Isis*, XLIX (1958), 378–390.

It is now just fifty years since the appearance of the first volume in the valiant programs of Pierre Duhem to rehabilitate the reputation of medieval science. His valuable work did not meet with the prompt recognition which it deserved, and even today so-called "popular" writers continue to proclaim that Galileo had no precursors in dynamics. On the other hand, those who did recognize the significance of Duhem's pioneer efforts were inclined to be overenthusiastic in accepting his judgments. Recent studies have given more balanced perspective with respect to claims of medieval anticipations of laws of dynamics, and it is the purpose of this paper to call attention to the need for a parallel reevaluation of ubiquitous judgments that the correct theory of the rainbow was discovered in the fourteenth century.

In 1814 Giambattista Venturi published a description of a manuscript copy, which he had rediscovered, of the *De iride* of Theodoric of Freiberg, a treatise composed around the years 1304 to 1310. In the first flush of excitement Venturi found in Theodoric's work a "complete explanation of the principal rainbow," "the explanation of the rainbow commonly attributed to De Dominis and to Descartes." The manuscript of Theodoric, overlooked for hundreds of years, did indeed include views on the rainbow of striking modernity, and overenthusiasm at that time is understandable and pardonable, but the necessary corrective reaction in this case has been overlong delayed. Only two years ago A.C. Crombie, author of by far the best and most extensive account of medieval work on the rainbow, described the explanation of Theodoric as "identical" with that of Descartes, although later on the same page he grudgingly admitted an "improvement" on the part of the latter. We shall here consider the chief points of resemblance and the more considerable points of difference between the Cartesian and Theodorican explanations. ...

Any doubt that Theodoric had a clear idea of the manner in which the two rainbows' arcs are produced is dispelled by the eminently clear diagrams which accompany his description. It is indeed true that he "traced with great accuracy the course of the rays which produce rainbows." But when scholars from Venturi to Crombie identify the explanation with that of Descartes, they have gone wide of the mark. Theodoric had explained the *mechanics* of the formation of the rainbow, but he was unable to account either for the size or for the shape of the bow. Rays following paths such as those Theodoric indicated should send light to the eye from many directions, and the problem of why only two narrow arcs (rather than the sky as a whole) are brightened remained unresolved. Theodoric was not unaware of the need for further explanation; but he was a better observer than theorizer, and we now see that mere observation scarcely could be expected to disclose the reasons for the magnitude of the bow. In place of these reasons Theodoric supplied such phrases as, "in the place determined by nature," or, "it is thus ordained by nature," to account for the limited range of effectiveness of the reflected and refracted rays. In the chapters where he came to grips with the question of the size and shape of the rainbow, Theodoric had recourse to the only quasi-quantitative theory which had yet been proposed—that of Aristotle. Neither a slavish imitator of Aristotelian views, nor yet a captious critic, Theodoric espoused an admirable suspension of judgment:

We say that one should teach that which the Philosopher said, for the authority of his philosophic doctrine and for the respect it deserves; and each one should interpret that which is said according to the same Philosopher, that one never should depart from that which is evident from the senses.

Nevertheless, and notwithstanding the fact that virtually all analyses of the work of Theodoric fail to mention the fact, dependence of his quantitative ideas on those of the Master is pronounced. The role of the beguiling meteorological hemisphere is accepted without question; and in the dozen-odd diagrams illustrating the formation of the two rainbows, the drops invariably are marshaled along the vertical great semicircle (of this hemisphere) at one end of which the sun is located. The prominence given by Theodoric to this illusory semicircle or "circle of altitude" (see figures 4 and 6) serves to indicate how much closer in quantitative theory he was to Aristotle than to Descartes. While one is justified in marveling at the modernity of Theodoric's qualitative explanation, one should not overlook his acquiescence in the incredible Aristotelian postulate that the drops producing the bow lie, together with the sun, on a sphere centered in the eye of the observer. Theodoric even went so far as to calculate, from an assumed value for the apparent size of the rainbow, the magnitude of the Aristotelian effective ratio. Although the admirable estimate of 42° for the radius of the bow was familar to many medieval scholars, Theodoric adopted, for reasons which are not clear, a value of 22°. From this hopelessly inaccurate radius he calculated for the effective ratio (i.e., for the ratio of the distances of the drop from the eye and the sun) a value of 60:118.

The theory of the effective ratio, we now see, was entirely misdirected, for it sought the quantitative clue in that seductive Helen of the rainbow, the meteorological hemisphere. Apparently Theodoric felt uneasy about fair Helen and the macrocosmic geometric explanation, for he tried systematically to link it to the microcosmic geometry of the raindrop. In so doing he proposed the only quasi-quantitative theory of the rainbow to appear in the long interval from Aristotle to Copernicus. So far as I am aware, no later commentator or historian has called attention to the nature or significance of this theory, despite the fact that it played a prominent role in the Theodorical explanation. Inasmuch as this represents the first effort to associate the size of the bow with the measures of angles relative to the raindrop itself, some account of it is certainly called for. Theodoric was led by his diagrams illustrating the formation of the primary rainbow to believe that the higher the position of the drop on the circle of altitude, the larger is the arc on the surface of the drop between the points at which the incident and emergent rays pierce the drop. Consequently he concluded that it is the magnitude of this arc—which we shall call the Theodorican arc—which governs the effectiveness of the rays reflected to the eye. The outermost part of the bow is red, he argued, because the Theodorican arc in this case is larger, and hence the reflected rays are stronger, than for rays striking drops at lower altitudes. The latter rays are weaker because of the smaller arc, and hence the colors are less bright, tending toward blue. For drops at still lower altitude the Theodorican arc is still smaller, and the reflected rays produce only a weak, albescent light, such as is seen within the rainbow arc. Above the red band, on the other hand, the arc is too great for rays to be reflected to the eye, he postulated, for reflection comes only at the proper altitude; and hence the region outside the bow is quite dark. For the secondary rainbow arc, on the other hand, the arc on the drop between the points of incidence and emergence of the rays he believed to vary *inversely* with the altitude of the drop on the circle of altitude. Hence the *lowest* portion of the secondary bow is red (the strongest and brightest color) and the uppermost part is blue (the weakest color). The region beyond the blue is illuminated only by a weak whitish light toward the zenith, for the critical arc is smaller; and within the red band no rays at all are transmitted to the eye, following the double internal reflection within the drops, because the Theodorican arc is too great. For the secondary bow as for the primary, therefore, it is the arc between the incident and emergent ray which determines the position of the bow.

The theory of Theodoric, apparently accounting for the properties of the two rainbow arcs, and presumably resolving the Aphrodisian paradox, possesses a high degree of plausibility; but it has two drawbacks. In the first place, it is incomplete, for Theodoric was unable to show how his arc should influence the quality of the reflection—why the colors begin to appear only when the Theodorican arc has attained a particular size, and why, when the arc has reached a certain critical magnitude, all reflection suddenly ceases. In other words, his theory failed to account quantitatively for the radius of the bow. In the second place, the theory, as we shall see below, is wrong. To correct his theory, two changes in particular must be made: (1) one must discard entirely the Aristotelian macro-cosmic circle of altitude; and (2) one must replace the Theodorican microcosmic arc between the points of incidence and emergence by the geometric angle between the incident and emergent rays. Theodoric, like other ancient and medieval scholars, had failed to observe that the drops producing the bow can be either at arm's length from the observer or miles away, *providing only that the appropriate angle is formed.* Not until the spell of the deceptive meteorological hemisphere was broken did mankind take the decisive step in the theory of the rainbow; and this was not a medieval achievement. ...

One might be inclined to agree with Sarton that Theodoric had gone just as far as the state of knowledge at the time permitted. It is generally held, in fact, that it was the discovery of the law of refraction which made possible the Cartesian theory of the rainbow. If this were so, it would be quite unreasonable to expect Theodoric to have taken the crucial quantitative step, and the word "failure" in our title would be unrealistic. It is easily demonstrated, however, that the Cartesian quantitative theory for the primary rainbow could well have been anticipated by Theodoric on the basis of the knowledge of refraction available in his own day. To show this, it will be necessary to describe very briefly the key discovery of Descartes.

Theodoric's attempt to find the quantitative clue to the rainbow was thwarted by his preconceptions about the role of the meteor-ological hemisphere. Descartes, although he had been brought up on the Aristotelian *Meteorologica*, abandoned the Master's hemi-sphere entirely and focused attention on the relationship of the solar rays to the drop alone. He repeated, apparently independently, the laboratory experiments of Theodoric, and noted again the rays which his medieval predecessor had found to be effective in the formation of the two rainbows. To determine the reason for the peculiar efficacy of these rays was, Descartes realized, "the prin-cipal difficulty." He therefore did what Theodoric failed to do—he *calculated* the paths of numerous rays through the raindrop. This is easily done if one knows, at least approximately, the relationship between the angles of incidence and refraction. Let O be the center of the raindrop (figure 7), I the point of incidence of the ray from the sun, R the point at the rear of the drop where the ray is reflected, E the point of emergence of the ray, and P the point of intersection of the incident and emergent rays (extended). The angle IPE, the apparent altitude of the rays reflected to the eye, is easily calculated in terms of the angles of incidence i and refraction r if one recalls that the angle ORI, which is equal to the angle of refraction, is measured by half the sum of the two remote interior angles of the

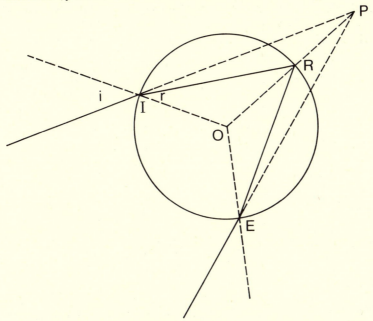

Fig. 7. Geometric construction for comparing Theodoric's and Descartes' studies of the "rainbow angle."

triangle RPI. If for the rainbow angle EPI one uses the symbol θ, and if one notes that angle RIP is equal to $i-r$, one can write the equation $r = \frac{1}{2}\theta + i - r$, from which one has $\theta = 4r - 2i$. Using this relationship, in conjunction with the law of refraction (between air and water), $3\sin i = 4\sin r$, Descartes drew up the following table (Table 1) illustrating the relationship between the angles i and θ. In this table he noted two facts of paramount importance: (1) for increasing angles of incidence, the angle θ at first increases, tends toward a maximum value of about $41°$, and then decreases; and (2) around this maximum angle there is a clustering of emergent rays leaving the drop in very nearly the same direction. The reason, then, that the sky is particularly bright along the primary rainbow band is that at this particular angle there are far more rays emerging from the drop than at smaller angles. Moreover, the darkness of the sky in Alexander's band above the primary rainbow arc is easily accounted for: Descartes noted that there are *no rays at all* returned to the eye at angles for which θ is greater than the maximum value of about 41°. Thus the shape of the principal bow, its radius, and the darkness of Alexander's band all are explained by Descartes' simple calculations. But could Theodoric have done the same in his day?

Table 1		Table 2	
$\sin i$	θ	i	θ
0.1	5° 40′	10°	12°
0.2	11° 19′	20°	22°
0.3	17° 56′	30°	30°
0.4	22° 30′	40°	36°
0.5	27° 52′	50°	40°
0.6	32° 56′	60°	42°
0.7	37° 26′	70°	42°
0.8	40° 44′	80°	40°
0.9	40° 57′	90°	36°
1	13° 40′		

The answer is easily given. The geometry needed for the discovery is so elementary that there can be no doubt but that this was well within Theodoric's power; but what about the law of refraction? Ptolemy had bequeathed the relationships between the angles of incidence and refraction for air to water shown in Table 3. The angles of incidence increase in simple arithmetic progression, and the angles of refraction increase in an arithmetic progression of second order, the first differences being in simple decreasing arithmetic progression (the first difference being $7\frac{1}{2}°$ and the common second difference being $\frac{1}{2}°$). It is easy to extrapolate to find that for an angle of incidence of 90° the angle of refraction should be 54°.

Table 3

i	10°	20°	30°	40°	50°	60°	70°	80°
r	8°	15½°	22½°	29°	35°	40½°	45½°	50°

This table was perpetuated by Witelo, with whose work Theodoric undoubtedly was familiar. Witelo modified the table slightly, giving 7° 45′ instead of 8° as the angle of refraction corresponding to an angle of incidence of 10°, but this does not alter the argument here. Had Theodoric carried through the reasoning of Descartes, using the Ptolemy-Witelo refraction table, he would have been led to the angles in Table 2. A glance at this table shows that the conclusions to which Descartes was led stand out even more sharply than in the Cartesian tabulation. The angle θ at first increases, reaches a maximum value of about 42°, and then decreases; and there is an obvious clustering of emergent rays at an angle of 42°, the very measure which Bacon and Witelo had given for the radius of the rainbow! That the Theodorican arc IE (figure 7) between the points of incidence and emergence does not play a decisive role becomes clear from the observation that this arc (which measures twice inscribed angle IRE) is four times the angle of refraction, and hence is given by the equation $IE = 4 \arcsin(\frac{3}{4} \sin i)$. The arc IE does not take on a maximum value within the range of Table 2, for it increases monotonically from almost 30° for $i = 10°$ to more than 194° for $i = 90°$. At the incidence $i = 65°$, close to the critical value for the rainbow, the Theodorican arc is somewhat over 171°. Rays returned to the eye therefore do not correspond to a unique value of the arc of Theodoric. From Table 2 one sees that light reaches the eye at an elevation of say 40° after incidence at either 50° or 80°, and the Theodorican arcs on the drops corresponding to these angles of incidence are respectively about 140° and 190°. Consequently the direct correspondence which Theodoric thought he saw between his arc and the colors in the rainbow turns out to be a chimera.

Theodoric failed to unravel the major mystery of the rainbow, even though it lay within his power to have done so. His theory is not that of Descartes; it is rather a half-way stage between the two chief systems of the rainbow, the Aristotelian and the Cartesian. ...

CHAPTER 6

STUDIES OF LOCAL MOTION

It is in the analysis and treatment of local motion that Aristotle's works are perhaps weakest, and it was in elaborating and criticizing his remarks that medieval scientists did some of their most impressive and fruitful work. The commentators of late Antiquity had attempted some clarification, and both the Muslims and the thirteenth-century Latins had dealt with the problem, but it was not until the early fourteenth century that significant progress was made. This progress resulted for the most part from the rigorous application of mathematics to the study of motion; the experimental side of the question was largely ignored. The scholars involved made many references to everyday experience, but they did not devise specific experiments the way that Roger Bacon, Witelo, and Theodoric of Freiberg had done.

In Aristotle's analysis, local motion was one of four possible species of motion, the others being change of substance, quality, or quantity. Change of place, or local motion, could be either natural or violent. Natural motion occurred when a body, having been removed from its "natural place," returned to it. There were two qualities of bodies which determined this motion: heaviness in the case of earth and to a lesser extent water, and lightness in the case of fire and to a lesser extent air. Violent motion occurred when some exterior force caused the body to move in some direction other than its natural inclination moved it. Such a force, in his view, could act only by maintaining permanent contact with the object it moved, and when this contact ceased, the violent motion should also cease.

Aristotle also held that natural motions (whether of heavy bodies downward or of light bodies upward) accelerated, but his description of this acceleration and his attempts to account for it were extremely inadequate. His Muslim commentator. Averroes, attributing human qualities to these bodies, interpreted Aristotle as meaning that as bodies neared their natural place, they became more desirous of reaching home as quickly as possible and so hastened their speed. But Aristotle also stated that velocity increased with the distance a body moved in natural motion.

In any case, he tended to ignore the factor of acceleration in his further analysis of natural motion and to treat it as being of constant velocity. The weight (or lightness) of a body and the resistance of the medium through which motion occurred were considered to be the crucial factors determining velocity, since the motion would be caused by the weight or lightness of the body and impeded by the resistance of the medium. He considered resistance an essential factor, claiming that without it motion would be instantaneous, which is impossible. In insisting on the necessity of considering resistance as necessary for motion, Aristotle, by his great authority, diverted many thinkers from pursuing more fruitful lines of investigation.

In violent motion, velocity was determined by the ratio between the force moving the body and the resistance of the medium. Aristotle was not consistent in his treatment of this question either, and although he said that there would be no motion unless the force were greater than the resistance, a rigorous application of the principles he enunciated would necessitate some motion for any values of these two factors.

It was especially in his *Physics* and *On the Heavens and the World* (*On the Heavens and the World* was always considered a single work in the Middle Ages, but in fact *On the World* was not written by Aristotle) that Aristotle treated the question of local motion, and it was largely in commentaries on these two works, especially the latter, that his remarks were questioned, expanded, and corrected throughout Antiquity and the Middle Ages. We know from the commentary on *The Heavens* by the sixth-century A.D. writer, Simplicius, what Hipparchus (one of the greatest of the ancient astronomers), Strato, and other unnamed ancient commentators had to say about local motion.

Hipparchus seems to have considered that constant contact between mover and moved was not necessary, but that a projectile receives a kind of residual force from its mover (or from that which prevented its motion), which is dissipated over a time. Thus if one throws an object straight up, it moves swiftly at first because the residual force imparted by the hand is great; it then slows down as this force weakens, begins to fall when the natural weight of the body (which, contrary to Aristotle, he seems to consider a constant) overcomes the residual force, and falls more and more rapidly as the residual force continues to weaken. The same argument applies to a heavy body dropped from a high place, in which case the force which originally prevented its fall confers upon it a resistance to its natural inclination to fall.

Another ancient scientist, Strato the Physicist, contributed significantly to the study of motion by two cases of acute reasoning on empirical evidence, in both of which he is concerned to relate velocity to distance of fall. In the first of these he notices that when water pours down from a high roof, it is continuous at the top but discontinuous near the ground. Strato reasoned that for this to occur, the water must traverse each successive unit of space more swiftly. Secondly, contradicting Aristotle's assertion that a thing moves more rapidly as it approaches its natural place, Strato noted that a stone dropped from about an inch made a barely perceptible impact on the ground (and thus was not moving swiftly), whereas if the same stone were dropped from a great height it would hit the ground with great force. From this he concluded that a freely falling body increases its velocity depending on the distance of its fall.

Simplicius mentions another theory, without naming its proponent, which held that a heavy body moves more quickly near the ground because the resistance of the air is less near the earth, since the height of a column of air becomes less as proximity to earth increases.

Simplicius' commentary was translated into Arabic very early and consequently influenced Arabic authors sooner than it did Latin writers. It was translated into Latin by William of Moerbeke in 1271 and from that time on was generally known in Latin Europe. During the last quarter of the thirteenth century Europeans gave more attention to the problem of local motion than they had earlier. Their works often took issue with Aristotle but remained largely within the Aristotelian frame of reference. They made some progress and foreshadowed many of the fourteenth-century developments, which indeed depended to some extent on them. In kinematics (i.e., the formal relationships of time, distance, and velocity) it was noticed that in the case of freely falling bodies, continuingly greater units of space are traversed in successive equal periods of time, and that velocity of fall increases with distance from the beginning of the fall rather than with proximity to natural place. In dynamics (the study of the actual forces involved in motion), there were several attempts to account for the continued motion of a projectile after it had lost contact with the mover and to explain the acceleration of a body in free fall. But these insights, important as they were, were not worked out, nor were either their consequences or the exact definition of their terms made explicit.

It was during the fourteenth century that the medieval study of local motion reached its height, especially at the universities of Oxford and Paris. It took place within the Aristotelian tradition, usually in the form of lectures (Commentaries or Questions) on Aristotle's *Physics* or *On the Heavens and the World*, but its net result was the destruction of the Aristotelian world view during the sixteenth and seventeenth centuries.

The first important studies of this sort were carried on at Merton College, Oxford by a group of natural scientists who were greatly interested in the application of mathematics to physics. The major figure among the Merton scholars was Thomas Bradwardine. In the year 1328 he composed a traetise *On the Proportions of the Speed of Motions* for the purpose of giving a clear mathematical account of the laws of motion and the relationship between force, resistance and velocity. His procedure was rigorously mathematical, beginning with a theoretical analysis of proportionality and then proceeding to deduce eight theorems from eight axioms. On the basis thus established, Bradwardine first disproved four widely held theories and then went on to construct his own views.

The following account is based upon, and in part quoted from, H. Lamar Crosby, Jr., ed., *Thomas of Bradwardine His Tractatus de Proportioninus* (Madison: The University of Wisconsin Press: © 1955 by the Regents of the University of Wisconsin). Because Bradwardine had no shorthand algebraic notation at his disposal and consequently expressed his formulas verbally, I have given Dr. Crosby's equivalent formulations in modern notation on the assumption that this will be more comprehensible to a modern reader. In this selection, V = velocity, F = motive force, R = resistance, k = some constant, n = any real positive number.

Bradwardine first stated his law of motion in his criticism of the third false theory, held by most medieval Aristotelians, that velocity is a direct proportional function of forces (i.e. $kV = F/R$). After objecting that this formula lacks generality, since it posits that either F or R must be considered as constant if two velocities are to be compared, he notes further that it leads to false consequents:

"1. The first false consequent is that any force, however small, can move any resistance, however large. For if $kV = F/R$, then $kV/2 = F/nR$. Since however large F may be, it will be possible to posit n such that nR will be larger than F, this theory is seen to be a violation of the universally accepted Aristotelian axiom that F must exceed R if there is to be any motion.

"2. The second false consequent is similar. For if $kV = F/R$, then because $kV/n = F/nR$, any *mobile* [moveable thing] can be moved by any force.

"3. Furthermore, sense experience teaches the opposite of this view, for we see, for example, that if one man can scarcely move a heavy rock, two men working together can move it much more than twice as rapidly. The same principle is illustrated in the case of clock weights: to double a weight may more than double the speed of descent. Therefore, Bradwardine concludes, $kV \neq F/R$.

"The remainder of his discussion of Theory III [that $kV = F/R$] is concerned with an attempt to show that Aristotle [and certain of his followers] could not have meant that $kV = F/R = D/T$ (D = distance, T = time), but that, instead, V varies as the proportion of F/R or D/T (i.e., $n^V = F/R$)." Having disposed of the erroneous theories, Bradwardine begins in Chapter III of his work to present his own views. "Now that these fogs of ignorance, these winds of demonstration, have been put to flight, it remains for the light of knowledge and of truth to shine forth. For true knowledge proposes a fifth theory which states that the proportion of the speeds of motions varies in accordance with the proportion of the power of the mover to the power of the thing moved." He assumes his function, $V = \log_n (F/R)$ or $n^V = F/R$, to be proven by the refutation of the four preceding theories. Of the remaining eleven theorems in this section, the most significant is the eighth—"No motion follows from either a proportion of equality [numerator and denominator are equal] or one of lesser inequality [denominator larger than numerator] between mover and moved"—by which he shows mathematically why no motion can result if force and resistance are equal or if resistance is greater than force; that is, if $F/R = 1/1$ or $1/1+$, then $V = 0$. The reason for this is (in Bradwardine's terms) that there is no proportion which expresses the relation of a proportion of equality to one of inequality; or in modern terminology, there is no exponential function of $1/1$ which will yield $1 + /1$. In other words the log of 1 to any base is 0.

Theorem IX proves by theorems I and VIII that if $F/R = 1 + /1$, then $V = +$. Theorem X, proved by theorems I and IX, states that for any value of V, proportions of F to R can be found which will yield $2V$ and $\frac{1}{2}V$. "Theorem X shows an important consequence of the $n^V = F/R$ formula. Since velocities vary as the proportion between their proportions of force to resistance, any given velocity may be doubled by squaring, or halved by extracting the square root of, the proportion associated with that velocity. Theorem X, therefore, presents the universal rule for the squaring and halving of velocities .... This theorem firmly establishes the power of the $n^V = F/R$ formula, not only to express any desired velocity by means of a proportion of forces, but to relate this velocity to any multiple or fraction of it, by means of another proportion (i.e., an exponential one)."

In arguing against an objection which might be brought against his views, Bradwardine made a distinction of fundamental importance between "quantitative" and "qualitative" velocity, and asserted that his study was concerned with the latter. "It is argued that if equal velocities are produced by equal proportions of forces, then, if a large quantity of earth bears the same proportion to a large quantity of air that a small quantity of earth bears to a small quantity of air, the velocities of the two bits of earth through their respective media should (by Bradwardine's account) be equal. But they cannot be, because the larger piece of earth, in traversing its medium in the same time that the smaller traverses its own medium, must go farther (because the medium is bigger in extent, or quantity), and in such an event the velocities would not be equal. "This dilemma thus raises the problem of relating force to distance, dynamic to kinematic functions, and brings out the ambiguity inherent in Aristotle's remark that the "weight" of a body is a factor of its velocity in free fall. Bradwardine offers, in solution of the above dilemma, a distinction between "qualitative" and "quantitative" proportionality as applied to motion. Qualitatively, the moving force bears the same proportion to any and all fractions of the impeding medium, and this qualitative proportion determines qualitative velocity. Quantitatively, however, the proportion is between the times of the two motions.

"Velocity as an 'instantaneous' quality of motion is thereby clearly distinguished from velocity as a simple function of time and distance. Needless to say, velocity of any sort must be thought of in terms of distance and time, but the distinction which Bradwardine here draws between quantitative and qualitative velocities is actually the distinction which may be rendered, in modern parlance, as that between $V = D/T$ and $V = dD/dT$. Thus, in Bradwardine's law, which is primarily dynamic, it must be understood that $V = dD/dT$ rather than simply D/T." This use of infinitesimals to express an instantaneous velocity considered as a quality of motion, and the unanswered question thus raised concerning the relationship between this quality of motion and the quantity of a motion possessing this quality, were enormously fruitful in subsequent studies of motion from the fourteenth to the seventeenth centuries.

"Bradwardine's *De proportionibus* represents... a significant advance in two very important aspects of the scientific enterprise: first, it moves forward with the task of developing mathematical formulae for the expression of physical laws whose entailed consequences do not contradict other generally accepted laws or observations—in other words the task of achieving self-consistency in physics through mathematics; second, through the introduction of mathematical analysis, it sets the stage for the quantitative measurement of physical processes and, hence, that typically modern physics which was to appear with Galileo's wedding of mathematics and experimental observation. Bradwardine used mathematics for the systematic and general expression of theory; Galileo used it for the systematic generalization of experimental observation."

Bradwardine's *Proportions* was not only a work of the first importance, considered in itself, but it was also extremely influential.
Within a very short time, it penetrated to Paris, eastern Europe and
northern Italy. At Merton College itself, Bradwardine was succeeded by a group of mathematical physicists known as the *Calculatores*.
Chief among them are Richard Swineshead, William Heytesbury,
and John Dumbleton. Among their major achievements were: the
Merton "mean speed" theorem, by which it was proved that in any
uniformly accelerated motion, the velocity reached at the mid-point
is equal to the average velocity of the entire movement; the equivalent of the correct formula (i.e., $S = \frac{1}{2}at^2$) in relating elapsed time
to distance traversed in a uniformly accelerated motion (this may
have been Dumbleton's interpretation of Bradwardine, but his
wording is ambiguous); and the use of line segments as primitive
graphs, visually representing arithmetic and logical procedures in
relating such terms as velocity and distance. Bradwardine's work
and those of the other Mertonians spread across Europe within a
few years' time and directly influenced nearly all subsequent discussions of dynamics.

The use of two dimensional (rather than straight line) graphs was
developed independently between 1346 and 1360 by an Italian,
Giovanni di Casali, and a Frenchman, Nicole Oresme. Casali was
first in time, but Oresme's treatment is much better. In a work called
The Configurations of Qualities he developed a method of describing
variations of qualities by means of geometric figures. Although his
system was designed generally to describe any kind of change, it was
also applied specifically to the problem of local motion. A case of
uniformly accelerated motion would be represented by marking off
equal units of time along the horizontal axis, which Oresme called
the base line, and velocities corresponding to these times along the
vertical, and then connecting the determined points by what he
called the summit line. The resultant figure would be a right triangle
whose area would represent the distance traversed. Other kinds of
qualitative change would be represented by other curves.

Mathematics, however, can only describe how things happen.
Why motion occurred was another question which occupied the
attention of many fourteenth-century writers as Aristotle's conten-

tion, that contact between mover and moved must constantly be maintained, began to break down under close scrutiny. Probably the most fruitful concept to come out of the discussions of this problem was that of "impetus," first fully developed in Latin Europe by John Buridan of Paris, although similar theories had been developed by several Arabic authors during the preceding three centuries. There had been several approaches to the impetus theory among the ancients, the Muslims, and the Latins, but Buridan's statement of it, while still containing a few ambiguities, is more comprehensive and precise than any which preceded him.

Buridan, born sometime before 1300 and dying in 1358, spent most of his adult life as a lecturer on the Arts faculty at Paris, expounding the books of Aristotle's natural philosophy. In the study of motion, his most important contribution was his doctrine of impetus. His problem was to explain why a body continues in motion after it loses contact with its original mover. After examining in detail and finding unsatisfactory the answers given by Aristotle and others, he proposes his own solution: that in addition to imparting motion to a body, the original moving force, whether gravity in the case of natural motion, or a human hand, a machine, or a bowstring in the case of violent motion, also imparts an impetus—a kind of force, he says, but not what we usually mean by force. The nature of this impetus is somewhat ambiguous. Since its primary function is to keep a body in motion, it must be some sort of force or cause of motion. But since it arises from motion itself it must also be, to some extent, a result of motion or a quality of motion. If Buridan had meant it only in the latter sense, it would be identical with the momentum of "Newtonian" physics, and in any case it is defined by the same factors: the more matter in a body and the greater its velocity, the greater the impetus. But as the following selections will make clear, he more often considered it as a force continuing to act on a body. If he had applied Bradwardine's law of motion—with which he was familiar—to his impetus theory, he would have concluded correctly that a constant force results in a constant acceleration rather than constant velocity, and thus the first way of understanding impetus (i.e., as a cause of motion) would have become unnecessary. But he did not and the ambiguity remained. An interesting aspect of Buridan's impetus is its permanence. It would apparently keep a body in motion forever if it were not corrupted by contrary forces, such as internal resistance, gravity, or resistance of the medium. In fact, Buridan twice applies it in this way, once in the example of the revolving mill wheel which would turn forever if the mill did not wear out and there were no resistance, and once in his suggestion that God may have given each of the heavenly bodies the impetus He desired at the time He created them, and they have been running on their own impetus ever since, because there is no resistance or other corrupting force in the heavens.

Buridan was extremely famous and influential in his own lifetime, and his works sometimes served as advanced textbooks for other masters at various European universities. However, his works were not published when printing largely ousted manuscript works in the late fifteenth century. Fortunately, however, Buridan had a pupil, Albert of Saxony, whose own works were for the most part rewritten versions of his teacher's, and Albert's works were published at Paris in 1516 and 1518. It was through this edition of Albert's works that the thought of Buridan was known to Galileo, on whom it clearly had a strong influence.

Selections from John Buridan, *Questions on the Heavens and the World*, translated from the Latin text of E. A. Moody, ed., *Johannis Buridani Quaestiones Super Libris Quattuor De Caelo et Mundo* (Cambridge, Mass., 1942, reprint New York, 1970), 176–181.

Book II, Question 12

In the twelfth place, it is asked whether natural motion ought to be swifter at the end than at the beginning.

1. It is argued that it should not be, because contrary instances are found in the motions of the heavens and of animals, which are nevertheless natural. There is also an instance in the motion of a heavy body in that case where there were a greater resistance at the end of its motion than at the beginning, as when a stone falls to earth first through the air and then through the water.

2. If we posit that the medium be uniform, it seems that the downward motion of a heavy body ought to be equally swift at the begin-

ning, middle, and end, because there is always the same moving body, the same moving force (that is, the same gravity), and a homogeneous medium and consequently a constant resistance. With all these factors remaining the same, the motion ought to have a constant velocity, since an increase in speed only comes about through a greater excess of moving force over resistance of the medium.

Aristotle says the opposite, and it is apparent to sense experience. For if a stone falls from the top of a house and hits someone who is only slightly below the roof, it will not hurt him. But if it continues to fall and hits a man standing on the ground, it will injure him because of its great velocity.

The fact that this is so should not be doubted because, as we have said, everyone perceives that the downward motion of a heavy body becomes increasingly swift if we posit that it falls through a uniform medium, because everyone perceives that the greater the distance a stone falls upon a man, the more seriously it will injure him.

But there is considerable doubt as to why this is so, and there are many different opinions about it. The Commentator [Averroes] in the second book of his commentary on this work makes some obscure remarks about it, saying that a heavy body moves more swiftly as it approaches the end of its fall because of its great desire to reach this place, and because of the heat caused by its motion. And from these words two opinions have sprung.

The first opinion was that motion causes heat, as is held in the second book of this work, and that therefore a heavy body falling rapidly through the air heats the air and consequently rarefies it; and the air, thus rarefied, is more easily divided and less resistant. Now if the resistance is diminished, it is reasonable that the motion should become swifter.

But these opinions are insufficient: First, because the air in summer is notably hotter than in winter, and nevertheless the same stone falling an equal distance in summer as in winter does not move notably faster in summer than in winter, nor does it strike with sensibly greater force. Also, the air is only heated by motion when it is previously moved and divided; and therefore, since it would resist a falling object before it were moved and divided, the air's resistance is not diminished as a result of its being heated. Furthermore, a man moves his hand just as swiftly as a stone falls at the beginning of its movement. This is apparent because hitting another person hurts him more than a falling stone would, even though the stone might be harder than the fist. And nevertheless a man does not sensibly heat up the air sufficiently that it would speed up to the extent that is apparent at the end of its motion.

Another opinion which has arisen from the words of the Commentator is this: Place is, in a way, a final cause of the located body as Aristotle hints and the Commentator explains in the fourth book of the *Physics*. And along with this, some people say that place is the cause moving the heavy body, as a magnet attracts iron. But whichever of these ways it occurred, it seems reasonable that the closer a heavy body approaches its natural place, the more swiftly it ought to move; because if place is the cause of movement, then it ought to move the heavy body more strongly when the two are close together—for an agent acts more strongly on what is near to it than on what is far away. And if place is nothing but the final cause, which a heavy body naturally desires and for the attainment of which it moves, nevertheless it seems reasonable that the natural desire for its end should increase when the end is nearer. And thus it seems completely reasonable that a heavy body moves more swiftly as it is closer to its natural place below. By falling continually it is brought closer to this place. Therefore by falling continually it ought to move more and more swiftly.

But this opinion cannot stand. ... It is against manifest experience, since you can lift the same stone near the earth just as easily as you could lift the same stone in a high place, such as the top of a tower, if the stone happened to be there. This would not be so if it had a stronger inclination toward its natural place when it was in a low place than in a high one. It is responded to this that indeed the stone does have a greater downward inclination when it is in a low place than when it is high, but not enough more that it might be perceived by sense. Such a response is not valid, because if the stone should fall continually from the top of the tower to the earth, two or three times as great a velocity or impact will be sensed near the ground than in a high point near the beginning of the motion. And therefore there should be two or three times as great a cause of the velocity. And thus it follows that that inclination which you posited to be neither sensible nor notable, is not the cause of so great an increase in velocity.

Furthermore, let the stone begin to fall to earth from a high place, and let another similar stone begin to fall to earth from a lower place. Then these stones, when they were one foot above the ground, ought to be moving at the same speed if increase in velocity only comes about as a result of nearness to natural place, since they are both the same distance from their natural place. And nevertheless it is clear to the senses that the one which fell from a high place would be moving much more rapidly that the one which fell from a low place, and the one would kill a man while the other would not hurt him.

Also, if a stone should fall from a very high place for a distance of ten feet and finding an obstacle there should come to rest, and a similar stone should fall from a low place, also for a distance of ten feet, all the way to earth, one of these motions will not be perceived to be swifter than the other, although one would be nearer to its natural place of earth than the other.

Therefore I conclude that the accelerated natural motion of heavy and light bodies does not result from greater propinquity to natural place, but from something else either directly applied to the moving body, or removed at a distance from it, or varied in proportion to the length of the motion. Nor is this similar to a magnet attracting iron, because if the iron is nearer the magnet, it immediately begins to move more swiftly than when it was farther away. But this is not so of a heavy body with respect to its natural place.

The third opinion was that the further a heavy body fell, the less air there was under it, and less air is able to offer less resistance. And if the resistance were diminished and the gravity moving the body remained the same, it follows that the heavy body ought to move more swiftly.

But this opinion falls into the same difficulties as the preceding one because, as has been said, if two completely similar heavy bodies begin to fall, one from a very high place and the other from a low place, such as ten feet from the ground, these heavy bodies move at the same speed at the beginning of their motion, notwithstanding the fact that one of them has much air underneath it and the other has little. And therefore in every case a greater velocity does not occur because of nearness to earth or having less air underneath, but from the fact that a moving body moves from a higher point and through a greater distance.

Besides, it is not true that the smaller amount of air in the aforementioned case would offer less resistance, because when a stone is near the earth there is still just as much lateral air as there was farther from earth, and therefore it is just as difficult for the divided air to yield and flee sideways as when the stone was farther from earth. Indeed, it is just as difficult, or more so, for the air to yield straight downwards when the stone is near the earth because the earth gets in the way, and it offers a greater resistance than does the air. Therefore this imagination [i.e., proposed solution] is not valid.

Therefore, now that we have dealt with these proposals, there remains one necessary way of considering the true cause of acceleration, so it seems to me. For I suppose that the natural gravity of such a stone remains always the same, before the motion, after the motion, and during the motion, and therefore the stone is found to be equally heavy after the motion as before. I also suppose that the resistance which results from the medium remains the same or quite similar, because, as I have said, it does not appear to me that the air which is lower and nearer the earth ought to offer less resistance than the higher air; but on the contrary, perhaps the higher air would offer less resistance because it is more rare. In the third place, I suppose that if the moving body is the same, and the total moving force is the same, and the resistance is also the same or similar, the motion will remain equally swift because the proportion of moving force to moved body and to resistance will remain the same. Then I add that in the downward motion of a heavy body, the motion does not remain equally swift but becomes continually swifter.

From these things, it is concluded that another moving force is added to the moved thing in addition to the natural gravity which moved it from the beginning and which always remains the same. Then, beyond this, I say that this other moving force is not place, which would attract the heavy body as a magnet attracts iron; nor is it another force existing in place originating either from the sky or from something else, because it would follow immediately that the same heavy body would begin to move more quickly from a low place than from a high place, and we know by experience that the opposite of this is true. And the consequence is apparent that a moving force acts more strongly when it is near than when it is far away, other things being equal, just as iron immediately begins to move more swiftly if it is near the magnet than if it is far away.

And from these considerations, it follows that it is necessary to imagine that a heavy body acquires motion to itself not only from its principal mover, i.e., gravity, but it also acquires to itself, along with that motion, a certain impetus, which has the power of moving the heavy body along with its permanent natural gravity. And because this impetus is acquired in common with motion, therefore the swifter the motion, the greater and stronger the impetus. Thus therefore a heavy body is moved at the beginning of its fall by its natural gravity alone, and therefore it moves slowly. Then it is moved both by its gravity and by the acquired impetus at the same time, and therefore it is moved more swiftly. And because the motion becomes swifter, therefore the impetus also becomes greater and stronger, and thus the heavy body is moved by its natural gravity and by this greater impetus at the same time; and thus again it is moved more swiftly, and thus it continues to accelerate to the end of its fall. And just as this impetus is acquired in common with motion, thus in common with motion it decreases or disappears as the motion itself decreases or disappears.

And you can try an experiment to illustrate this: If you make a large and very heavy mill wheel rotate swiftly, and then you cease to move it, it will still keep on moving for some time because of its acquired impetus. Indeed, you cannot make it stop right away. But because of the resistance which results from the gravity of the mill, the impetus would continually diminish until the mill ceased to turn. And perhaps, if the mill should last forever without any diminution or change, and there were no other resistance to corrupt the impetus, the mill would move forever because of its perpetual impetus.

And thus someone might imagine that it would not be necessary to posit intelligences to move the heavenly bodies, because Scripture does not say anywhere that they ought to be posited. For it could be said that when God created the celestial spheres, He began to move each one of them as He wished. And they are still moved by the impetus He gave them, because this impetus is neither corrupted nor diminished, since it has no resistance.

And you ought to note that some people have called this impetus "accidental gravity," and this name is not inappropriate. Whence, this seems to be in agreement with Aristotle and the Commentator in the first book of this work, where they say that the gravity would be infinite if a heavy body were moved infinitely, because the more it is moved, the more swiftly it is moved, and the more swiftly it is moved the greater is its gravity. Therefore, if this is true, a heavy body, in moving, should acquire to itself a continually greater gravity. And this acquired gravity is not of the same rationale and nature as the first natural gravity, which always remains even when motion ceases, whereas the acquired gravity does not. And all these things I have said will be more apparently true and necessary when we investigate the violent motions of projectiles and other things.

Book III, Question 2

Finally, it is asked whether a thrown stone or an arrow shot from a bow, after it loses contact with the projecting force, if continual movement is caused by an intrinsic or extrinsic principle.

1. It is argued in the first place that it is not moved by an intrinsic principle, because everyone concedes such motions to be violent. ...Therefore, such a motion is caused by an extrinsic principle and not by an intrinsic one.

2. Furthermore, in the second book of the *Physics* it is said that things which are moved in some way other than natural motion—and such projectiles are of this sort—have no innate (that is, intrinsic) principle of change.

3. Besides, the matter of a stone cannot cause its motion, because it has no activity. Nor can the form or gravity of the stone move it in this way—that is, upward or horizontally—but on the contrary they both incline it in the opposite direction, i.e., downward. There-

fore, it does not appear that there is anything intrinsic which might move the stone.

4. Again, Aristotle in book IV of the *Physics* says that such a projectile is moved either by *antiparistasis* [Buridan explains this term shortly] or because the air driving it from behind pushes it more rapidly than the lateral force moves it toward its natural place. And therefore he says that such a projectile would not move in a vacuum because none of these conditions would be found in a vacuum, and for this reason he decides that the projectile is moved either by the air behind it or by the surrounding air. And he seems to express the same meaning in the eighth book of the *Physics* and in the third book of this work. For he says in the third book of this work that the air, since it is naturally both heavy and light, moves quickly and easily both upward and downward, and when it is impelled upward it retains that part of its nature for a time because of its natural lightness, and thus it also retains its motion downward because of its gravity if it is impelled downward. And thus he finally posits that the air, driven upward with the projectile, moves the projectile upward; and if the heavy body is moved downward naturally, the air still moves this body onward because of its natural gravity and makes the motion swifter. Aristotle seems expressly to hold this opinion, and the Commentator explains all of his words as though this is what they mean.

Nevertheless, the opposite position is argued, because in the natural downward motion of a heavy body, the air seems to resist it, and it does not seem to offer any less resistance to its upward motion. And that which resists both the motion and the moved thing cannot at the same time move that thing.

We inquire furthermore what thing would move the air when the original propelling force ceased, especially if the projectile were moving horizontally. For if you should say that the air moves itself, then I could say the same thing about the projectile [and hence the air's motion would be unnecessary]. If you should say that a certain force is impressed upon the air by the original propelling force, I can likewise say the same thing about the projectile. And if you should say that it is moved by its own gravity or levity, this does not seem reasonable, because gravity and levity only incline a body to move naturally either upward or downward; and thus just as much doubt remains about the motion of the air after the original propelling force has ceased as there is about the motion of the projectile.

Furthermore, it seems a remarkable thing that air, which is so easily divisible, would support a ten pound stone, such as are hurled by machines, for so long a time.

Aristotle mentions two major opinions concerning this question, both of which agree in this, that a thrown stone or an arrow shot from a bow would be moved by the air after they had left the original projecting force; because he supposes that every motion results from another motor so that nothing ever moves unless something else should move it. Now, the bow, or hand, or whatever, does not move it any more after it has lost contact with it; but on the contrary, if these were annihilated immediately, the stone or the arrow would nevertheless move the same distance. And it appears that there is nothing else in the vicinity which might move the stone or arrow except the air, because the actual mover ought to be in immediate contact with the moved thing, and nothing but the air seems thus to be present along with the stone or arrow. Therefore Aristotle concludes that the stone or arrow is thus moved by the air. And then he mentions two ways in which this might be done. The first is what he calls "by *antiparistasis*," that is, that when a stone is thrown, it leaves the place in which it was before, and then nature, since she abhors a vacuum, quickly sends the air behind it in to fill up the empty space. And this rapidly moving air, behind and touching the projectile, drives it yet farther ahead, and so on to the end of its flight.

But Aristotle does not hold this position, nor ought it in any way to be held. In the first place the problem would immediately recur about a thing moving circularly, such as a hoop or a mill wheel, not evacuating any place, for it happens that they continue to move for a long time after the original projecting force has ceased—that is, the man rolling a hoop or turning a mill wheel—and then one must still decide what moves the hoop or mill wheel. And this problem cannot be solved by the aforementioned *antiparistasis*.

Here is another obvious proof from experience: If a ship full of hay is pulled rapidly by horses against the current of the river, and then the horses cease to pull it, the ship will still move for a time, nor could it be stopped immediately. Therefore, if you should say that the air behind the boat has so much strength that it could continue to move the boat against the current of the water, it would follow that the air behind the boat would completely mat down and blow over the reeds and grasses behind the boat. Nevertheless, this whole explanation appears to be false, and therefore the air does not push the boat. We could appeal to many other facts of experience, but there is no need, since Aristotle does not hold this opinion.

There is another explanation which Aristotle and the Commentator both seem to hold, namely that the impelling force moves the adjoining air along with the stone or the arrow, and that the air, which is well suited by nature to move, is moved rapidly, and by its rapid motion it moves the projectile a certain distance. And when it is asked by what the air is moved, the Commentator responds that it is moved by a principle intrinsic to itself, that is, by its own natural gravity or levity, so that in whatever direction it is driven, is has the nature of retaining that motion for a certain time as the result of its natural gravity or levity.

But it still appears to me that this opinion in no way saves the appearances. In the first place, concerning the hoop and mill wheel, if you should say that the surrounding air moves so great a weight circularly after a man ceases to move it, I would object. Because if you should take a rag and wipe the contiguous air away from the wheel, you will not stop the wheel in this way. And if the wheel were enclosed between boards, ... nevertheless the wheel would continue to revolve for just as long as if it were not enclosed, and nevertheless it does not seem that so little air which would be between the wheel and the enclosure could move it for so long a time and so swiftly. Similarly, I object concerning the boat: If it were wrapped in cloth and driven swiftly, and then suddenly the cloth—and the surrounding air with it—were removed, the boat would continue to move just as fast.

And here is something else to wonder about. If the air which I set in motion when I throw a stone can move the stone, why will it be that if I blow the air at you as swiftly as I can without the stone, you can hardly feel it? For you ought to feel it strongly if it were able to bear a large rock.

And furthermore, why is it that you cannot throw a feather as far as five feet? For if, when the air is set in motion, it moves the projectile, it ought to move the feather farther and more easily than it moves a heavy rock.

Therefore, since these appearances and many others cannot be saved by that opinion, I rather think that the mover impresses on the moved thing not only motion, but along with it a certain impetus or some force or other quality—not the kind of force we usually mean by that name—which impetus has the nature of moving that thing on which it is impressed, just as a magnet impresses on iron a certain force moving the iron to the magnet. And the more swift the motion, the more intense the impetus will be. And this impetus in a rock or arrow is continually diminished by the resistance contrary to itself until it is no longer able to move the projectile. If you find some way of saving the opinion of Aristotle and the appearances at the same time, I shall gladly adopt that view.

Selections from Marshall Clagett, *The Science of Mechanics in the Middle Ages* (Madison: The University of Wisconsin Press; © 1959 by the Regents of the University of Wisconsin), pp. 552–556, 570.

Of Buridan's successors the most interesting and important one is Nicole Oresme, who altered Buridan's view of the nature of impetus. In his Latin commentary of Aristotle's *De caelo*, Oresme held that impetus arises from an initial acceleration and then acts to accelerate further the speed... Some time later in his French commentary on the *De caelo*, he says much the same sort of thing. The acquisition of impetus comes from acceleration. As Miss Maier has pointed out, the question of how the initial acceleration takes place is not clearly answered by Oresme. As to the nature of Oresme's impetus and its generation, he says, "it is a certain quality of the second species...; it is generated by the motor by means of motion, just as it would be said of heat, when motion is the cause of heat.... It is corrupted by the retardation of motion because for its conservation speed or acceleration is required." Thus Oresme's impetus differed from Buridan's in two major respects. (1) It was no longer simply a function of velocity but apparently of acceleration as well. (2) It was no longer considered to be of permanent nature. Hence, it is not surprising that Oresme does not explain the movement of the heavens in terms of impetus.

It should be remarked that Oresme later applied his doctrine of the impetus produced from acceleration to projectile motion. In such motion the propellant introduces an initial acceleration into a body which causes an impetus which in turn moves the body after there is no longer any contact with the projector. The impetus then by its action further accelerates the movement of the projectile until it is sufficiently weakened by resistance, at which time deceleration occurs.

One of the most interesting parts of Oresme's exposition concerning falling bodies is that where he suggests that if we pierced the earth so that a body could fall to its center, the body would acquire *impetuosité* which would carry it beyond the center of the earth; and so, rather than coming to rest immediately at the center of the earth, it would oscillate about the center of earth at gradually decreasing distances until it finally came to rest (see Doc. 9. 3)...

Perhaps even more interesting than Oresme's account of the impetus theory is his apparent acceptance of the correct description of the acceleration of falling bodies in his Latin commentary on the *De caelo* of Aristotle:

That something is "continually accelerated" can be understood in two ways. In one way thus: An addition of velocity takes place by equal parts, or equivalently. For example, in this hour it is moved with some velocity, and in the second twice as quickly, and in the third three times as quickly, etc. In the same way such an addition can be made in proportional parts of the time. And in this way infinite velocity would accordingly follow if one proceeded to infinity, because any given velocity would be exceeded in this way of increase. In the second way, addition of velocity can be imagined not by equal parts but by continually proportional and smaller parts, so that if now the velocity were on one degree, next it would be $1 + \frac{1}{2}$ degrees, and then $1 + \frac{1}{2} + \frac{1}{4}$ degrees, etc. By this way a double velocity (i.e., one of 2 degrees) would never be exceeded, even though one proceeded to infinity. Now as for the question at hand of the acceleration of falling bodies, velocity in the motion of a heavy body increases in the first way and not the second.

It should be observed that this description is still somewhat ambiguous.... Now it would appear that Oresme believed that, in the natural motion of a heavy body, the first method of arithmetical increase in equal periods of time was the proper description, although he might mean that the velocity increase was arithmetical with some other than equal divisions of the time period. It should be clear that Oresme has not... considered in this passage the infinitesimal aspects of the description of uniform acceleration evident in the Merton treatment of "motion uniformly difform," although, of course, Oresme was quite familiar with that kinematic activity....

So far as I know, the first statement of free fall with the infinitesimal implications of the Merton discussions explicitly applied is found in the *Questiones super octo libros Physicorum Aristotelis* published by the Spaniard Domingo de Soto in 1555, where he says:

Movement uniformly nonuniform as to time is nonuniform in such a manner that if it is divided according to time (i.e., according to before and after), the middle point of any part at all exceeds in velocity the least velocity of that part by the same proportion that the mean is exceeded by the greatest velocity of that part. This species of movement belongs properly to things which are moved naturally and to projectiles. For when a body falls through a uniform medium, it is moved more quickly in the end than in the beginning. On the other hand, the movement of projectiles upward is less quick in the end than in the beginning. And so the first is uniformly increased, while the second is uniformly decreased.

Soto then goes on to say, just as the Merton College kinematicists had said, that a movement uniformly accelerated is measured or denominated with respect to the space traversed in a given time by its mean velocity. The example (a thought experiment only) which Soto gives is of a body falling through an hour which accelerated uniformly from zero degree of velocity to a velocity of eight. Then it would traverse, he says, just as much space as another body moving uniformly through the hour with a speed of four. Soto, then, has applied the mean law of Merton College to falling bodies....

At this point, I should like merely to contrast with the medieval discussions the treatment of falling bodies given by Galileo in his *Two New Sciences* of 1638.... He is interested only in the *kinematic* description of falling bodies.... Now he shows by means of the inclined plane experiment given below that a deduction from the mean speed theorem, namely $S \propto t^2$, applies to the fall of bodies. His procedure is this. Accept as a fundamental principle that uniform acceleration is defined as $V \propto t$. From this deduce that $S \propto t^2$. Reason that the case of free fall is like the case of balls rolling down inclined planes. Now show by rolling balls down an inclined plane that the relationship $S \propto t^2$ holds. Conclude, then, that it also holds for free fall.

Regardless of how well he performed his experiments and what data came out of those experiments, Galileo's treatment was certainly the starting point of modern investigations of the problem of the acceleration of falling bodies.

Nicole Oresme, *On the Book of the Heavens and the World of Aristotle.* Translated by Marshall Clagett from the Old French edition of A. D. Menut and A. J. Denomy, in *Mediaeval Studies*, Vol. 3 (1941), 230–31.

Book I, Chapter 17... And when he says that the weight is greater just as the velocity is greater, we are not to understand by "weight" a natural quality which inclines downward. For if a stone of one pound should descend from a high place so that the movement was swifter in the end than at the beginning, the stone still would have no more natural weight at the one time than at the other. But we ought to understand by this "weight" which increases in descent an accidental quality which is caused by the compulsion of the increase in the velocity, as I have said on another occasion in the seventh book of the *Physics.* And this quality can be called "impetuosity." And it is not weight properly speaking because if a passage were pierced from here to the center of the earth or still further, and something heavy were to descend in this passage or hole, when it arrived at the center it would pass on further and ascend by means of this accidental and acquired quality, and then it would descend again, going and coming several times in the way that a weight which hangs from a beam by a long cord swings back and forth. And so this impetuosity is not properly weight, since it causes ascent. And such a quality exists in every movement—natural and violent—as long as the velocity increases, the movement of the heavens being excepted. And such a quality is the cause of the continued movement of projected things when they are no longer in contact with the hand or the instrument of projection....

CHAPTER 7

ASTRONOMY

Although astronomy was probably the most persistently studied of all the scientific disciplines during the Middle Ages, since it was indispensable in setting the dates of movable feast days, it was nevertheless nearly always considered merely as a practical device for saving the appearances and getting results. Even in the early Middle Ages, scholars were aware of a variety of cosmological schemes devised by the ancients and passed on in the handbooks, so there was never a question of one specific cosmology commanding the adherence of all. Gerbert in the tenth century had adopted the scheme of Pliny, slightly modified by that of Plato, on the grounds of simplicity and clarity, although he was fully aware of other possibilities.

The translations of the twelfth century further complicated the matter by making available to the Latins two additional schemes, those of Aristotle and Ptolemy, both of which were widely influential though mutually incompatible. Aristotle's cosmology was based mainly on philosophical considerations and was most persuasively argued, but it was of little use in solving particular problems such as establishing the date of Easter, predicting new moons and conjunctions and oppositions of planets, determining the solstices and equinoxes. Ptolemy's was a fully elaborated mathematical construction giving the orbits of the individual planets. It did a magnificent job of saving the appearances, but it did not accord with or bother much about problems of celestial mechanics. So one system made sense but did not work; the other worked but did not make sense. By far the majority of medieval writers on astronomy, if they faced the question at all, explicitly stated that in astronomy one could not know the way things really were but must be satisfied with saving the appearances. Very few argued that one particular scheme corresponded to the actual structure of the cosmos.

The Middle Ages made less progress in the study of astronomy than in almost any other scientific discipline, but by the middle of the fourteenth century a breakthrough seemed immanent. Both Buridan and Oresme argued well for the relativity of the perception

of motion and made attacks on portions of the generally received
celestial mechanics of their day. Still, both ended up retaining the
view that the earth was probably motionless in (or near) the center
of the universe.

There was no pressing need for a change; the Ptolemaic system
gave satisfactory results. But ignorance of the way things really are
rankles, and men began to seek more seriously the "truth" about
the structure of the cosmos. The most spectacular and influential
revision was that of Copernicus (1543), which despite its myriad
shortcomings is correctly considered to be the seminal work of modern astronomy. That Copernicus could or would have written his
book without the preceding works of the fourteenth-century astronomers seems highly unlikely.

We have already noted Buridan's daring suggestion that God may
have given the heavenly bodies an original impetus and then let
them alone. This willingness to break new ground is also evident in
his treatment of one of the problems crucial not only to astronomy,
but to the entire physical structure which the Middle Ages inherited
from Antiquity—that is, the question of the earth's daily rotation.
The system of Heraclides of Pontus was known (sometimes in
garbled form) through the ancient handbooks. This system had assumed that the earth was in the center of the universe and rotated
daily on its axis, that the planets Mercury and Venus revolved
around the sun, and that the sun and the other planets revolved
around the earth. There were always a few astronomical writers
who accepted all or part of Heraclides' system, but the overwhelm-
ing majority, whether Aristotelian or Ptolemaic, considered the earth
to be at rest and the sphere of the fixed stars, as well as the planets,
to be in daily motion around the earth. The question which Buridan
addressed himself to was whether the earth was always at rest in the
center of the universe. In considering this problem, he devised (as
well as borrowed) some compelling arguments showing that perceived motion is relative and that the astronomical appearances
would be the same whether it was the earth or the heavens that
moved. There was one experiment, however, which seemed to him
to be convincing against the earth's daily rotation, namely that if
one shot an arrow straight up, it would come down at or near the

spot from which it was shot, rather than considerably to the west. And so, because of what he considered to be convincing experimental evidence, he decided that the earth, after all, was at rest. But his successor, Nicole Oresme, took up and expanded this discussion, and in the process explained away the problem of the arrow.

Selections from John Buridan, *Questions on the Heavens and the World*, translated from the Latin text of E. A. Moody, ed., *Johannis Buridani Quaestiones super Quattuor Libris De Caelo et Mundo* (New York, 1942), 226–233.

Book II, Question 22

It is asked consequently whether the earth is always at rest in the middle of the world.

1. And it is argued that it is not, because there is some motion which is natural to every natural body. Therefore the earth either moves naturally or at least it can move naturally. And if it can move naturally, then it ought to move sometimes, because it would be improper to say that its natural potency is eternally frustrated and never becomes actual.

2. Furthermore, the earth is spherical in shape, and a spherical figure has a certain aptitude for circular motion. Now, as we have just said about its potency, I also say now, that a natural aptitude should not remain eternally unrealized.

3. Also, Aristotle says that there ought to be some simple movement which is natural to every simple body....

4. Also, the ancients have argued that the more noble element ought to occupy the more noble place; and fire is more noble than earth; therefore fire ought to have a more noble place. But the nobler place, and the place in which the located thing can best be preserved, is the center. For this reason, the king is accustomed to place himself in the middle of his kingdom, for he is safer there since his enemies cannot get at him quickly.

Aristotle argues the opposite point of view in this place; and first he also posits this consequence: The heaven always moves circularly; therefore the earth is always at rest in the center.

This question is difficult, for in the first place there is considerable doubt as to whether the earth is exactly in the center of the world so that its center and the center of the world coincide. There is another serious doubt as to whether the earth as a whole might sometimes be moved in a straight line, since we do not doubt that many of its parts are often moved, for this is apparent to us through our senses. And still another difficult problem concerning the soundness of Aristotle's consequence—namely that if the heavens necessarily move circularly, then it is also necessary for the earth to be at rest in the center. The fourth problem is whether, by positing that the earth moves circularly around its center and its own poles, all of our appearances can be saved. And we shall talk about this last problem now.

It should be known that many people have held it as probable that it does not contradict the appearances for the earth to move circularly in the aforesaid manner and for it to make one revolution from west to east in a natural day, with reference to a designated point on its surface. And then it is necessary to posit that the stellar sphere is at rest; and then day and night would be caused by this motion of the earth in such a way that the motion of the earth would be a daily motion. And I can give you an example illustrating this. If someone is moved in a boat and imagines himself to be at rest, and if he should see another boat which was really at rest, it would appear to him that the other boat is moving, because his eye would be in exactly the same relationship to the other boat whether his own boat was moving and the other was at rest, or the other way around. And thus we might also posit that the sphere of the sun is completely at rest and that the earth spins, carrying us along with it. Since we would nevertheless imagine ourselves to be at rest, just as the man on the swiftly moving boat did not perceive either his motion or that of his boat, therefore it is certain that to us the sun would rise and set in the same manner as it does when it moves and we are at rest.

But it is nevertheless true that if things were as we have just suggested, one would have to concede that the spheres of the planets move, because otherwise the planets would not change their positions relative to the fixed stars. And therefore this opinion imagines that each planetary sphere moves just like the earth, that is from west to east, but because the earth has a smaller circle it accomplishes its rotation in a shorter time, and the moon in a shorter time than the sun, and so on for all the planets, so that the earth completes its rotation in a natural day, the moon in a month, the sun in a year, etc.

And it is undoubtedly true that if things were the way this view posits, everything in the heavens would appear to us just as it does now. We ought to know too that those wishing to uphold this opinion, perhaps for the sake of argument, put forth certain arguments in its favor.

The first is that the sky does not need the earth or other lower bodies in order to acquire anything for itself, but on the contrary the earth needs to acquire for itself influences from the sky. Now it is more reasonable that that which needs something should move in order to acquire it than that which does not need anything. ...

The third argument is that more noble conditions ought to be attributed to heavenly bodies, and especially to the highest sphere. But it is more noble and more perfect to be at rest than to move. Therefore the highest sphere ought to be at rest. ...

The last argument is that just as it is better to save the appearances with fewer things than with more, if this is possible, so it is also better to save them in the easier way rather than the more difficult. It is easier to move a small thing than a large one. Therefore, it is better to say that the earth, which is very small, is moved rapidly and the highest sphere is at rest, than the contrary.

Nevertheless, this opinion ought not to be held, in the first place because it is against the authority of Aristotle and all the astronomers. But they reply that authority does not demonstrate, and that it is sufficient for astronomers to devise a means of saving the appearances, whether their scheme is true in fact or not. The appearances are saved by both means, and so they can adopt the one which pleases them best.

Others argue by many appeals to experience. One of these is that the stars appear to our senses to move from east to west. But they solve this problem by saying that things would appear the same if the stars were at rest and the earth were moving from west to east.

Another appearance is that local motion causes heat, and therefore both we and the earth should become extremely hot as the result of so fast a motion. But they say that motion only causes heat by the friction of bodies, by rubbing, or by scattering. And none of these things happens here because the air, water, and earth all move together.

But the last appearance which Aristotle notes is more demonstrative in this matter, namely that an arrow shot straight upward from a bow falls to earth on the same spot from which it was shot; and this would not be so if the earth were turning very swiftly. Rather, before the arrow fell back, the part of the earth from which it was shot would be one league away. But still they attempt a reply, saying that this happens because the air moved with the earth carries the arrow with it, although the arrow appears to us to have only a straight line motion because it is carried along with us, and therefore we do not perceive the motion by which it was carried along with the air. But this evasion does not suffice, because the violent impetus of the arrow in rising would resist the horizontal motion of the air in such a way that the arrow would not move horizontally as much as the air would. For example, if the air were moved by a strong wind, an arrow shot straight upward would not move so far horizontally as the air, but it would move somewhat.

Nicole Oresme, in his own commentary on Aristotle's *On the Heaven and the World*, took up Buridan's question again. Portions of the two discussions are very similar, except that Oresme's arguments are often fuller and more cogent, and he solves the problem which had led Buridan to abandon the idea of the earth's rotation—i.e., the case of the arrow shot straight upward—by the brilliant suggestion that the arrow would share the earth's movement and that its apparent rectilinear flight was in fact a compound motion of which we only perceive one component. To be true to philosophy, Oresme should have accepted the fact of the earth's daily rotation. Whether his concluding remarks were ironic, whether he genuinely wished to assert the supremacy of irrational faith over reason, whether his recoil was from innate conservatism or from a shattering insight into the far-reaching consequences of such a position, I do not know. But his final published position was that, all convincing reasons to the contrary, the earth stood still.

Selections from Nicole Oresme, *On the Book of the Heavens and the World of Aristotle*, translated by Marshall Clagett, *The Science of Mechanics in the Middle Ages*, 600–606, from the Old French edition of A. D. Menut and A. J. Demony in *Mediaeval Studies*, IV (1942), 270–279.

1. Book II, Chapter 25. Afterwards he (Aristotle) sets forth another opinion.

Text: And some say that the earth is at the center of the universe and that it revolves and moves circularly around the pole established for this, just as is written in the book of Plato called the *Timaeus*.

Gloss: This was the opinion of one called Heraclitus Ponticus who proposed that the earth is moved circularly and that the heavens are at rest. Aristotle does not here refute these opinions, perhaps because it seemed to him that they have little root in appearance and are pretty well refuted elsewhere in philosophy and astronomy.

But it seems to me, subject to correction, that one could support well and give luster to the last opinion, namely, that the earth, and not the heavens, is moved with a daily movement. Firstly, I wish to state that one could not demonstrate the contrary by any experience. Secondly I will show that the contrary cannot be demonstrated by reasoning. And thirdly, I will put forth reasons in support of it (i.e., the diurnal rotation of the earth).

2. As for the first point, one experience commonly cited in support of the daily motion of the heaven is the following: We see with our senses the sun and moon and many stars rise and set from day to day, and some stars turn around the arctic pole. This could not be except by the movement of the heavens, as was demonstrated in chapter 26. Thus the argument runs, the heaven is moved with a diurnal movement. Another experience cited is this: if the earth is so moved, it makes a complete turn in a single natural day. Therefore, we and the trees and the houses are moved toward the east very swiftly, and so it should seem to us that the air and the wind blow continuously and very strongly from the east, much as it does against a quarrel shot, only very much more strongly. But the contrary appears by experience. The third experience is that which Ptolemy advances: if a person were on a ship moved rapidly eastward and an arrow were shot directly upward, it ought not to fall on the ship but a good distance westward away from the ship. Simi-

larly, if the earth is moved so very swiftly in turning from west to east, and it has been posited that one throws a stone directly above, then it ought to fall, not on the place from which it left, but rather a good distance to the west. But in fact the contrary is apparent.

It seems to me that by using what I shall say regarding these experiences, one could respond to all the other experiences which might be adduced in this matter...

3. ... Again, I make the supposition that local motion can be sensibly perceived only in so far as one may perceive one body to be differently disposed with respect to another. In support of this I give the following illustration: If a person is in one ship called *a* which is moved very carefully i.e., without pitching or rolling—either rapidly or slowly—and this person sees nothing except another ship called *b*, which is moved in every respect in the same manner as *a* in which he is situated, I say that it will seem to this person that neither ship is moving. And if *a* is at rest and *b* is moved, it will appear and seem to him that *b* is moved. On the other hand, if *a* is moved and *b* at rest, it will appear to him as before that *a* is at rest and that *b* is moved. And thus, if *a* were at rest for an hour and *b* were moved, and then immediately in the following hour the situation were reversed, namely, that *a* were moved and *b* were at rest, this person on *a* could not perceive this mutation or change. Rather it would continually seem to him that *b* was moved; and this is apparent by experience. The reason for this is because these two bodies, *a* and *b*, are continually changing their dispositions with respect to each other in the same manner throughout when *a* is moved and *b* is at rest as they were conversely when *b* is moved and *a* is at rest. This is apparent in the fourth book of *The Perspective* of Witelo, who says that one can perceive movement only in such a way as one perceives one body to be differently disposed in comparison with another. I say, then, that if the upper of the two parts of the universe mentioned above should today move with a diurnal movement, just as it is, and the lower part should not, and tomorrow the contrary should prevail, namely that the lower should be moved with a diurnal movement while the upper (i.e., the heavens) should not, we could not perceive this change in any way, but everything would seem to be the same today and tomorrow. It would seem to

us continually that the part where we are situated was at rest and that the other part was always moved, just as it seems to a person who is in a moving ship that the trees outside are moved. Similarly, if a person were in the heavens and it were posited that they were moved with a diurnal movement, and furthermore that this man who is transported with the heaven could see the earth clearly and distinctly and its mountains, valleys, rivers, towns, and chateaux, it would seem to him that the earth was moved with a diurnal movement, just as it seems to us who are on the earth that the heavens move. Similarly, if the earth and not the heavens were moved with a diurnal movement, it would seem to us that the earth was at rest and the heavens were moved. This can be imagined easily by anyone with good intelligence. From this reasoning is evident the response to the first experience, since one could say that the sun and the stars appear thus to set and rise and the heavens to turn as the result of the movement of the earth and its elements where we are situated.

4. To the second experience, according to this opinion, the response is this: Not only is the earth so moved diurnally, but with it the water and the air, as was said, in such a way that the water and the lower air are moved differently than they are by winds and other causes. It is like this situation: If air were enclosed in a moving ship, it would seem to the person situated in this air that it was not moved.

5. To the third experience, which seems more effective, i.e., the experience concerning the arrow or stone projected upward etc., one would say that the arrow is trajected upwards and simultaneously with this trajection it is moved eastward very swiftly with the air through which it passes and with all the mass of the lower part of the universe mentioned above, it all being moved with a diurnal movement. For this reason the arrow returns to the place on the earth from which it left. This appears possible by analogy: if a person were on a ship moving toward the east very swiftly without his being aware of the movement, and he drew his hand downward, describing a straight line against the mast of the ship, it would seem to him that his hand was moved with rectilinear movement only. According to this opinion of the diurnal rotation of the earth, it seems to us in the same way that the arrow descends or ascends in a straight line...

... In support of this position, consider the following: If a man in that ship were going westward less swiftly than the ship was going eastward, it would seem to him that he was approaching the east, when actually he would be moving toward the west. Similarly, in the case put forth above, all the movements would seem to be as if the earth were at rest.

Also, in order to make clear the response to the third experience, I wish to add a natural example verified by Aristotle to the artificial example already given. It posits in the upper region of the air a portion of pure fire called *a*. This latter is of such a degree of lightness that it mounts to its highest possible point *b* near the concave surface of the heavens. I say that just as with the arrow in the case posited above, there would result in this case of the fire that the movement of *a* is composed of rectilinear movement, and, in part, of circular movement, because the region of the air and the sphere of fire through which *a* passes are moved, according to Aristotle, with circular movement. Thus if it were not so moved, *a* would ascend rectilinearly in the path *ab*, but because *b* is meanwhile moved to point *c* by the circular daily movement, it is apparent that *a* in ascending describes the line *ac* and the movement of *a* is composed of a rectilinear and a circular movement. So also would be the movement of the arrow, as was said. Such composition or mixture of movements was spoken of in the third chapter of the first book of the *De caelo*. ... I conclude then that one could not by any experience whatsoever demonstrate that the heavens and not the earth are moved with diurnal movement.

6. As to the second point relative to the rational demonstration of the diurnal movement of the heavens, I first note the following: It seems to me that this rational demonstration proceeds from these arguments which follow and to which I shall respond in such a fashion that, using the same reasoning, one could respond to all other arguments pertaining to it. ...

Again, if the heavens were not moved with diurnal movement, all astronomy would be false and a great part of natural philosophy where one supposes throughout this movement of the heavens.

Also, this seems to be against the Holy Scripture which says [in Eccles. I: 5–6]: "The sun riseth, and goeth down, and returneth to his place: and there rising again, maketh his round by the south, and turneth again to the north; the spirit goeth forward surveying all places round about, and returneth to his circuits" (Douay translation of Vulgate). And so it is written of the earth that God made it immobile: "For God created the orb of the earth, which will not be moved."

Also, the Scriptures say that the sun was halted in the time of Joshua (see Josh. 10:12–14) and that it returned (i.e., turned back) in the time of King Ezechias (Hezekiah; see Isa. 38:8; II Kings 20:11 [Vulgate, IV Kings 20:11]). If the earth were moved and not the heavens, as was said, such an arrestment would have been a returning and the returning of which it speaks would rather have been an arrestment. And this is against what the Scriptures state....

7. To the fifth argument, where it is said that if the heavens would not make a rotation from day to day, all astronomy would be false, etc., I answer this is not so because all aspects, conjunctions, oppositions, constellations, figures, and influences of the heavens would be completely just as they are, as is evident clearly from what was said in response to the first experience. The tables of the movements and all other books would be just as true as they are, except that in regard to the daily movement one would say of it that it is in the heavens "apparently" but in the earth "actually". There is no effect which follows from the one assumption more than from the other. Apropos of this is the statement of Aristotle in the sixteenth [actually the eighth chapter] chapter of the second book of the *De caelo*, namely, that the sun appears to us to turn and the stars to sparkle and twinkle because, he says, it makes no difference whether the thing one sees is moved or the sight is moved. Also one would say apropos of this matter of diurnal rotation that our sight is moved with diurnal rotation.

8. To the sixth argument concerning the Holy Scripture which says that the sun revolves, etc., one would say of it that it is in this part simply conforming to the manner of common human speech, just as is done in several places, e.g., where it is written that God is "repentant" and that he is "angry" and "pacified" and other such things which are not just as they sound. Also appropriate to our question, we read that God covers the heavens with clouds—"who

covereth the heavens with clouds" (Ps. 146:8)—and yet in reality the heavens cover the clouds. Thus one would say that according to appearances the heavens and not the earth are moved with a diurnal motion, while in actuality the contrary is true. Concerning the earth, one would say it is not moved *from* its place in actuality, nor *in* its place apparently, but that it is moved *in* its place actually. To the seventh argument, one would answer in just about the same way, that according to appearances in the time of Joshua the sun was arrested and in the time of Ezechias it returned, but actually the earth was arrested in the time of Joshua and advanced or speeded up its movement in the time of Ezechias. It would make no difference as to effect whichever opinion was followed. This latter opinion supporting the diurnal rotation of the earth seems to be more reasonable than the former, as we shall make clear later.

As to the third main point of this gloss, I wish to put forth persuasions or reasons by which it would appear that the earth is moved as was indicated....

9. Again, all philosophers say that something done by several or largescale operations which can be done by less or smaller operations is done for nought. And Aristotle says in the eighth chapter [actually the fourth chapter] of the first book that God and Nature do not do anything for nought. But if it is so that the heavens are moved with a diurnal movement, it becomes necessary to posit in the principal bodies of the world and in the heavens two contrary kinds of movement, one of an east-to-west kind and others of the opposite kind, as has been said often. With this theory of the diurnal movement of the heavens it becomes necessary to posit an excessively great speed. This will become clear to one who considers thoughtfully the height or distance of the heaven, its magnitude, and that of its circuit; for if such a circuit is completed in one day, one could not imagine nor conceive of how the swiftness of the heaven is so marvelously and excessively great. It is so unthinkable and inestimable. Since all the effects which we see can be accomplished, and all the appearances saved, by substituting for this diurnal movement of the heavens a small operation, i.e., the diurnal movement of the earth, which is very small in comparison with the heavens, and since this can be done without making the number of

necessary operations so diverse and outrageously great, it follows that if the heaven rather than the earth is moved then God and Nature would have made and ordained things for nought. But this is not fitting, as was said.

Again, when it has been posited that the whole heavens are moved with daily movement and in addition that the eighth sphere is moved with another movement, as the astronomers posit, it becomes necessary, according to them, to assume a ninth sphere which is moved with a daily movement only. But when it has been posited that the earth is moved as was said, the eighth sphere is moved with a single slow movement and thus it is not necessary with this theory to dream up or imagine a ninth natural sphere, invisible and without stars; for God and Nature would not have made this sphere for nought, since all things can be as they are by using another method. ...

10. It is apparent, then, how one cannot demonstrate by any experience whatever that the heavens are moved with daily movement, because, regardless of whether it has been posited that the heavens and not the earth are so moved or that the earth and not the heavens is moved, if an observer is in the heavens and he sees the earth clearly, it (the earth) would seem to be moved. The sight is not deceived in this, because it senses or sees nothing except that there is movement. But if it is relative to any such body, this judgment is made by the senses from inside that body, just as he [Witelo] stated in *The Perspective*; and such senses are often deceived in such cases, just as was said before concerning the person who is in the moving ship. Afterwards it was demonstrated how it cannot be concluded by reasoning that the heavens are so moved. Thirdly, reasons have been put forth in support of the contrary position, namely that the heavens are not so moved. Yet, nevertheless, everyone holds, and I believe, that they (the heavens), and not the earth, are so moved, for "God created the orb of the earth, which will not be moved" (Ps. 92:1), notwithstanding the arguments to the contrary. This is because they are "persuasions" which do not make the conclusions evident. But having considered everything which has been said, one could by this believe that the earth and not the heavens is so moved, and there is no evidence to the contrary. Nevertheless, this seems prima facie as much, or more, against natural reason as

are all or several articles of our faith. Thus, that which I have said by way of diversion in this manner can be valuable to refute and check those who would impugn our faith by argument....

There is no way to prove that Copernicus read either Buridan or Oresme, but both men were philosophers of preeminence, whose works were used in universities throughout Europe. The likelihood that Copernicus read them is much higher than the likelihood that he did not, since they had reached Cracow, where Copernicus went to school, and were common in northern Italy, where he pursued advanced studies. In any case, the extreme similarity of the arguments used by Copernicus on the question of the earth's rotation to those put forth by both Buridan and Oresme increases the probability significantly. Marshall Clagett has pointed out the most important of these similarities: 1. arguments against the huge velocity of rotation which would be required of a daily rotating heaven; 2. the air rotates along with the earth; 3. the example of the man in the moving boat to illustrate the relativity of perceived motion; 4. falling bodies combine rectilinear and circular motion; 5. the nobler parts of the universe—i.e., the heavens—should be at rest.

The commentaries of these two fourteenth-century French philosophers were widely read, and their arguments provided a good part of the context in which discussions of astronomy took place during the fifteenth century. They had not taken the new paths they had pointed out, but their work made it possible for others to do so.

CHAPTER 8

THE FRINGES OF SCIENCE

The science of the Middle Ages was not completely devoid of magical elements, although these are much less prominent than is usually supposed. It was especially in the sciences (or pseudo-sciences) of alchemy and astrology that the magical outlook predominated— that is, cause and effect relationships outside the realm of nature were considered to be operative. Then as now there was a large amount of sheer superstition, and both written magic lore and char-latan "magicians" abounded. What we are interested in in this chapter is the extent to which magical ideas influenced the world view of serious scientists.

There is a deep-seated affinity for magical thought in the human mind, and probably no man is completely exempt from it. But it is possible for one to be afraid of the dark, follow an unchanging ritual in setting up lab equipment, and wear a St. Christopher medal and still be entirely rational and naturalistic in his scientific work. It is also true that much science and philosophy have consisted of an attempt to rationalize folk superstition. Furthermore there are many areas where science and magic are difficult to distinguish. The planets do quite obviously exercise some power upon the earth. The light and heat of the sun, the motions of the sun and moon and their effects upon the seasons and the tides are clearly respectable objects of scientific investigation. Why then not go on to the other planets? Where does one stop? There is a very thin line too between number theory and numerology, between mathematics and number mysticism. Quite often the most important distinction between magic and science is the attitude of the author. The two are seldom wholly different. What makes Newton's universal gravitation less magical than Bacon's multiplication of species? Both assume an unexplained universal power, capable of mathematical description, operating at a distance.

In the Middle Ages, at least from the twelfth century, there were three primary types of attitudes toward magic among educated people. The first denied it completely and maintained a consistently rationalistic explanation of physical phenomena. A second accepted most of the magic lore available. The third was basically naturalistic but contaminated to a greater or lesser degree by magical ideas. The third was the most prevalent, but it is important to realize that the first type also existed.

Before the twelfth century, although there was considerable popular superstition in Europe, there was strictly speaking very little alchemy or astrology, simple because the complicated body of teaching on which they were based was not yet accessible to the Latins. By the end of the century, however, large amounts of such material had been translated from the ancient texts, almost wholly from Arabic. The *Tetrabiblos* of the great astronomer Ptolemy was prominent among the astrological works. This work is good evidence of the fact that it is quite possible for one man to be a first-rate scientist and still accept much supernatural teaching. One of the founders of modern astronomy, Johannes Kepler, was a practicing astrologer who publicly affirmed his belief in witches. There were also many Arabic works by such authors as Jabir, Rasis, Alkindi, Albumasar, Costa ben Luca, pseudo-Aristotle, and pseudo-Avicenna. Probably the most influential among works of this sort was a group of writings on astrology and alchemy attributed to Hermes Tresmegistos—the Thrice Great Hermes—a conflation of the Greek god Hermes and the Egyptian god Toth. Actually written by a number of different authors over a considerable span of time (between 150 B.C. and 150 A.D.), they possessed great authority because of their supposedly divine origin but not all Arabic authors accepted their doctrines. Avicenna, the principal explicator of Aristotle for the Latins before 1230 A.D., was consistently hostile to the entire alchemical doctrine. Many Arabic authors opposed the more extreme claims of the astrologers, nor did such books meet with wholehearted acceptance by Latin authors. Throughout the Middle Ages there was an ambivalent attitude toward them, although as the Middle Ages came to a close and the modern period began, acceptance became more general.

The literature on astrology is voluminous and for the most part easily understood. But it is otherwise with alchemy. Secrecy of doctrine was an important tenet of the alchemists, and this led them to disguise their writings as allegories, giving enigmatic names to their

reagents, and generally striving for obscurity. There are many alchemical "recipe books" and a fair number of theoretical works on
alchemy, but since they are largely incomprehensible and extremely
dull at that, we have not included any strictly alchemical selections
except for Michael Scot's brief recipe for making gold from copper.
We have, however, tried to indicate fairly what the attitudes toward
it were.

Many Latin scientists of the twelfth century were intoxicated by
the newly-acquired treasures of ancient and Arabic wisdom, and
many accepted uncritically everything they contained. An Englishman, Daniel of Morley, wrote a book *On the Natures of Things
Above and Below*, which shows the influence of these teachings on
his own thought. Daniel had begun his education in England and
had then gone to Paris. Annoyed by the dominance of law at Paris,
he went to Spain to continue his studies of philosophy, and there
he heard the great translator, Gerard of Cremona, lecture on
Ptolemy's *Almagest* and argue for the influence of the stars on the
lower world. He returned to England with many books of Arabic
wisdom. His only extant work was written in reply to the questions
of Bishop John of Norwich concerning what he had learned while
in Spain. Unlike Adelard of Bath's *Natural Questions*, Daniel's work
is based on an extensive first-hand acquaintance with Arabic writings
and probably gives a good picture of the views and learning of the
northern European avant garde during the last quarter of the twelfth
century. As in most works of this period, its world view is an original
compound of elements found in the traditional Latin sources and
the newly-acquired books of Greek and Arabic learning. The first
part is concerned with the lower world; the second, with the higher.
Although the entire treatise is extremely interesting, our selection
is limited to the concluding portion of Book II.

Selections from Daniel of Morley, *Liber de naturis inferiorum et superi-
orum*, translated from the Latin text of Karl Sudhoff, "*Daniels von Morley
Liber de naturis inferiorum et superiorum,*" *Archiv für die Geschichte der
Naturwissenschaften und der Technik*, VIII (1917), 31–38.

Since, of the whole multitude of stars, the seven planets are more
efficacious on the lower world than the rest, it will be sufficient to

treat of them. These planets move from one zodiacal sign to another, they rise and set over the lower world, and they bring about changes in it by their motions. For example, the power of the sun is manifest in many things, such as the heliotrope. Metals also attest to the effect of the sun; its power is apparent in the precious stone jasper and also in certain kinds of pearls. Even the uneducated do not doubt that the sun works changes on the lower world by means of the seasons of the year. For we see plants and trees in winter dry up because they are deprived of their natural heat and the benefit of their nourishment, and lose their leaves. But when spring comes, because they experience the benefit of heat, they are soon revived, thanks to this nourishment, and they begin to put forth leaves and to produce flowers and fruit, each according to its own kind. Whence the Philosopher [Aristotle] says in his book *On Generation and Corruption*, "The sun causes generation by coming near and decay by moving away." ... But the nature of the laurel is disobedient to the effect of the sun. For this effect to occur, two things are necessary, namely the nature of the acting body and the nature of the body acted upon. If the lower body does not cooperate, the effect of the higher will be destroyed or impeded.

The power of the moon is attested to by the menstrual cycle of women. The tides of the sea also affirm this. Their principal cause consists in the coming together of three things, namely the nature of the place, the condition of the water, and the motion of the moon. Of all the authors I have read, Albumasar treats of this in the most convincing fashion. And even more to the point, the power of the heaven is extremely great in medicine, so that the moon claims completely for herself the critical days, by which the variation of illness is understood; whence it pleased Hippocrates and Galen and many other physicians to compute the critical days according to the motion and the different mansions of the moon. Therefore, whoever damns astrology also necessarily destroys physics. He who is ignorant of the causes of things cannot easily administer his daily affairs. But the astrologer foresees the cause, what ought to be remedied, and why and when, so that the physician may then put this knowledge to good use. And I recall other effects of the moon which should not be passed over. For there are many kinds of things which increase when the moon is waxing and decrease when it is waning. Thus, when the moon is waxing, the humors in the bodies of animals become more abundant, but they thin out when it is waning. For whatever is cold and moist in animals, such as milk, the cerebrum and the medulla, becomes more abundant when the moon waxes but thins out when it wanes. And no less than this, the albumen of an egg conceived and layed in the first half of a lunation is more abundant. ...

Those who deny the power and efficacy of the heavens have the impudent madness to belittle the teachings of a science before they have learned them. Whence some men hold astrology in hatred from the name alone. But if they would give their attention to how many things of great usefulness and value come forth from it, they would never belittle it except from envy. Concerning its importance, according to what wise men say, it has eight parts: the science of judgments (judicial astrology); the science of medicine; the science of nigromancy according to physics; the science of agriculture; the science of illusions; the science of alchemy, which is the science of the transformation of metals into other species; the science of images, which is treated in that great and universal *Book of Venus* by Thoz the Greek; and the science of mirrors, which is broader and more inclusive than the others, as Aristotle makes clear in his book on the burning glass.

The utility of astrology ought to be considered no less carefully. For the astrologer, since he foresees events that are going to happen, can prevent or avoid the harmful ones, such as civil war, famine, earthquakes, fires, floods, and plagues of both men and beasts. And if he could not avoid these completely, nevertheless, having forseen the event, he would make much better provision for it than would the ignorant, who, taken by surprise, would be cut down by sheer terror. ... [Daniel now goes on to name each of the planets in order and to explain the terms equinox, axis, pole, the celestial circles, the five regions of the earth, the zodiac, eclipses, and the seasons. He cites several Arabic sources and criticizes Martianus Capella and all the other Latins for not properly understanding the theory of eccentric orbits.]

It now remains to treat of the twelve signs, each one of which occupies thirty degrees, and thus the whole zodiac, as it is considered by the physicist, has 360 degrees. The names of the signs are these: Aries, Taurus, Gemini, Cancer, Leo, Virgo, Libra, Scorpio, Saggitarius, Capricorn, Aquarius, Pisces. Each one of the planets has it own 'dignities' in these signs and can be either higher or lower. The 'houses' of Saturn are Capricorn and Aquarius, of Jupiter Saggitarius and Pisces, of Mars Aries and Scorpio, of Venus Taurus and Libra, of Mercury Gemini and Virgo, of the sun Leo, and the moon's house is Cancer. Aries acts especially on fire, Taurus on earth, Gemini on air, and Cancer on water. Going back to the beginning, Leo also corresponds to fire, Virgo to earth, Libra to air, and Scorpio to water. Starting over once more, Saggitarius corresponds to fire, Capricorn to earth, Aquarius to air, Pisces to water. There are therefore three fiery signs, three earthy, three airy, and three watery. The fiery signs are Aries, Saggitarius, Leo; the earthy are Taurus, Virgo, and Capricorn; the airy are Gemini, Libra, and Aquarius; and the watery are Cancer, Scorpio, and Pisces. Although these signs are not actually fiery, earthy, airy, and watery [since Daniel denies that the heavens are made of the four elements], they are nevertheless called by these names because when the sun is in them it heats us greatly (fiery) makes things cold and dry (earthy), etc. ...

We must not pass over the fact that the Arabs hold the powers of these signs in such veneration that they have divided the body of man himself according to their various powers; and in the first place they say that Aries rules the head and face in the human body, Taurus the neck and throat. They assign to Gemini the shoulders, arms, and hands. Cancer holds the breast with the lungs the esophagus and spleen with the ribs. Leo, according to them, holds the bottom of the stomach, which they call *merin*, and also the heart, the liver, and the flanks with the back. Virgo is allotted the belly with the intestines, Libra the navel and from the pubic hair to the loins, Scorpio the private parts from the loins to the buttocks and thigh, Capricorn the knees and their nerves, Aquarius the shin bones and the ankles, Pisces the feet and their nerves. Thus they also give the spleen to Saturn, since it is melancholy, the liver to Jupiter, the bladder with the bile to Mars, the heart to the sun, the shape to Venus, the mouth and tongue to Mercury, corpulence to the moon. Having arranged things thus, they affirm that slowness comes from Saturn, temperance from Jupiter, anger from Mars, the power of domination from the sun, wisdom and eloquence from Mercury, pleasure from Venus, and abundance of moisture from the moon, since it is the mother of moisture. From these seven planets, the seven days take their names. Therefore, the first day of the week is the sun's, the second the moon's, the third Mars', the fourth Mercury's, the fifth Jupiter's, the sixth Venus', the seventh Saturn's. The reasons why the days do not come in the same order as the planets requires a long explanation, and so let us pass over it here.

Now that I have enumerated the opinions of the astrologers, it should be known that when the aforementioned Arabs treat of the constellations, they call the planets Lords of the Nativities—whence also the saturnine from Saturn, the jovial from Jupiter (Jove), and so on with the others. Therefore, according to them, whoever is born under Saturn, since Saturn is dark, harsh, and heavy, will be melancholy, greedy, wandering to far-away and cold places, holding one thing in his mouth and another in his heart, desiring evil, envious, sorrowful, etc. But Jupiter, since it is a regal star, prosperous, sweet, and temperate, makes kings religious, wise, honest, and moderate in anger; makes legislators obedient, men of vision, truthful, constant, and generous; makes noblemen wealthy, lovers of women and pleasing to them. ... But Mars, since it is sharp, bitter and severe, makes kings violent, inhuman, fierce, and perverse; makes judges unjust, makes knights treacherous, lovers of battle, plunderers, oppressors, thieves, brigands, ... irascible, contemptuous, and haughty. The sun, a regal star, light and eye of the world, makes men handsome, lively, commanding, regal, ... religious, wise, ... rich, worthy, ... aiding the good and casting down the evil. Benevolent Venus, which is also called the star of mothers, makes a man generous, agreeable and affable with women, sexually active, debauched, given to drink and a gambler, desiring gold, silver, musical instruments, pleasures and good times, admiring beautiful shapes and pictures, and visiting temples with frequent veneration. But Mercury, since its nature is promiscuous and ready to be mixed with everything, concords with all the signs and planets, and this star makes clever negotiators, doctors of arts, mathematicians, geometers, astrologers, ... scribes, historians, men capable of every art, fleeing pleasures, ... perverse with the perverse, benign with the benign, quick with solutions, full of eloquence, businessmen and merchants, seeking and losing riches, giving away what they possess and possessing little. The moon, since it is benevolent, and is also called the star of the sun, makes princes and dukes the defenders of law, lifegiving, leaders, conciliators, protectors of mothers and sisters.

Therefore, these dominating influences of the stars, just as all never come together in one body at the same time, thus they do not come forth from one simple star, but they are collected from the conjunction of many. For the malevolence of Saturn can be mitigated by the power of any benevolent star, but it can also be augmented by a malevolent one....

Such views did not go unchallenged. We have already noted in Adelard of Bath a healthy distrust of magic at the very beginning of the twelfth century. About 1160 a scientist named Marius, probably a teacher at Salerno in southern Italy, wrote a remarkable treatise on the elements in which he held that all of physical nature was composed of a universal substance, which was differentiated by pairs of the four basic qualities (hot, cold, moist, and dry) into the four elements. These elements combined according to fixed ratios to produce the innumerable compound bodies of this world. Each element lost its actual identity in the compound but remained itself potentially, so that every compound could be resolved into its elements. Marius made up a table of the 145 different ways the four elements could combine and held that there were innumerable variations within each of the 145 types of compounds. In his treatment of the formation of metals (which was the special province of alchemy), he was completely naturalistic, holding that all metals are made of sulphur and mercury (the traditional doctrine) of varying degrees of purity, heated in the earth's interior by entrapped heat. The intensity of the heat, the length of time it operated, and the purity of the sulphur and mercury distinguished one metal from another. It is impossible, he said, to change one metal into another, although the colors can sometimes be altered. He also went out of

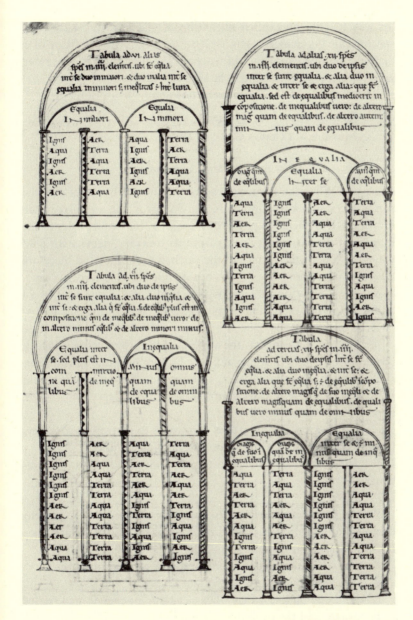

Fig. 8. One of Marius' tables showing how the elements combine. From British Museum MS Cotton Galba E. IV, folio 195ᵛ.

his way to offer an explicit refutation of magic, denying that any species can be changed. His treatise is written in the form of a dialogue between student and teacher. The first book is concerned with the elements themselves, the second with the compound bodies which are made of them. Our selections come from Book II.

Selections from Marius *On the Elements* (*De elementis*), translated from British Museum MS Cotton Galba E. IV.

1. *That only bodies act on other bodies*

Teacher. When the surface of the earth is heated by the heat of the sun in the summer, the moisture and vapors which are mixed with bits of earth are heated up. But in the winter, the heat recedes to the interior of the earth as if fleeing from its contrary, namely the cold which surrounds the surface of the earth at that time; and therefore the earth is cold on the surface, but its interior is hot at that time.

Student. How do you mean that heat flees? Since heat is an accident and not separated from substance, will you be able to discover it by itself?

T. Indeed, I mean that the vapors which are mixed with earth flee downward, and thus they take the heat in them into the interior.

S. That is enough concerning this question. But I wish you would point out to me how each contrary flees its contrary, which you apparently meant to say above.

T. Truly, every contrary tries and works to destroy its contrary, just as cold tries to destroy heat, and humidity dryness.

S. Isn't heat one of the qualities of bodies, and isn't it an accident?

T. It is.

S. And how can one quality so abhor another that it tries to demolish and destroy it, since a quality has no discretion nor does it know what it ought to do?

T. Do you know that sleeping and waking are contraries?

S. Yes, I know this.

T. And do you know that when sleep approaches, a man ceases to be awake, and the other way around?

S. I know this too.

T. Thus it should be understood about heat and cold, moisture and dryness.

2. *How a compound is formed*

T. Thus I wish you to understand, my dear pupil, that when the four elements make up any body, none of them is found actually in the composite body except insofar as it is there potentially. For example, in a fruit composed of the four elements, none of them is found there actually, but only potentially. For if fire were actually in the fruit, that fruit would undoubtedly be found to be hot. And if air were actually in the apple itself, the fire existing in it actually would destroy that air, and the fruit would not be durable. But we know that a fruit lasts for many months. If, however, water were actually in the fruit, it would either certainly flow out, or it would be turned into vapor by the force of the fire existing in it actually. But if earth were in the fruit actually, its own heaviness would clearly be the cause of its destruction. Another argument: if the four elements were actually in the fruit, one of them would not be able to exist for an hour with another. For it has been established that fire flees from water, and conversely, as has been established above in this book. Therefore, it is clear that a fruit can only exist if each one of the elements in it were changed into something other than it was before it became a fruit. But since fire would oppose water in the composition of an apple, fire operates on water and water undergoes its operation. Also in the same way water operates on fire, and fire undergoes the operation of water. And each of them tries to alter the other from its own nature, since when fire operates on water it heats it up and dries it out. And conversely when water operates on fire it imparts cold and moisture. And when in the same way air opposes earth in the composition of the fruit, both of them struggle to drive the other from their own nature, and each of them operates on the other and undergoes the operation of the other. When earth operates on air, it bestows solidity and dryness on it. But when air operates on earth, it makes it hot and moist. For this reason, therefore, there is made from the four elements another mixture, another temperament, and another complexion than existed before, since each one of them, from that state of actuality in which it existed earlier, is changed into something else. Nor will any of them be found in the fruit in that previous state of actuality, since each one is changed.

3. *How metals are formed*

[This is part of a larger discussion of the formation of minerals in general. Marius says that they all contain all four elements, and that they are formed from predominantly earth matter by entrapped heat in the earth's interior.]

T. It should be noted that all six kinds of metals are composed in the earth of quicksilver and sulphur, namely gold, silver, copper, iron, lead, and tin. This fact can also be proved from this, that when these are liquefied in a fire, they are similar to quicksilver. Therefore, gold is made from pure quicksilver mixed with pure sulphur heated in the earth's interior for a long time. And both because of the long time taken in the cooking of gold and because its material was pure, its parts were pressed closely together until it became solid and heavy. Therefore it neither rots inside the earth nor can it easily be burned by fire. Silver, however, is made from pure quicksilver mixed with a small amount of red sulphur for a short time. And because a small amount of red sulphur was present, therefore it did not have the same reddish-yellow color as gold. Copper was made from impure quicksilver mixed with a very dirty and somewhat dense red sulphur, and it was cooked for a very long time—even longer than gold—at an extremely high heat. And because it underwent a great force of combustion, it therefore has much redness in it. Also, its body was so loosened up that the vapor of vinegar is able to enter its body and give it a greenish color, which is called 'flower of brass.' Iron was made from quicksilver mixed with sulphur of a color halfway between red and white, and it was cooked for a long time. It was cooked even longer than copper by a moderate heat, and from this is derived its great solidity; nor does it become fluid from the force of fire as copper does, since it was cooked for a long time. And because a moderate heat was present, its body did not become loose like copper. And because the sulphur contained only a small amount of redness, therefore if iron should lie undisturbed for a long time, it becomes rusty and takes on a reddish color. Tin was made from pure quicksilver mixed with pure white sulphur. But it is cooked for a short time. For if in cooking it the heat were small and the time were great, it would be turned into silver. Lead was made from coarse quicksilver mixed with coarse sulphur which was white with just a little red. And that it was made from white sulphur is clear, because when it is placed near vinegar, it takes on a white color. But that the sulphur was slightly reddish is clear when lead burns, for then it is of a reddish color.

4. *That species cannot be changed*

S. Now I ask whether or not there is anything in addition to sensible and transitive motion which might to any extent be added to a plant, by which something different from the plant or animal might be brought about....

T. Indeed, nothing can be added to a plant except motion and rest. If you should add rest, when the motion ceases, it will be like a mineral. But if you add motion, it will either be the motion of doing something, of decaying, of growing larger or smaller, or even of changing into something else. If the motion of doing something is added, decay necessarily follows, for that from which anything was made now ceases to be what it was. If, however, the motion of decay were added, it is clear to everyone that it will decay. If the motion of growing larger or smaller were added, the plant will move as before, only more or less. But if the motion of changing into something else were added, there is no doubt that it ought to be changed into an animal or mineral. But this cannot be, for the power of this motion cannot change one species into another, but only one quality into another. Green can become more red or yellow, as we see in apples and pears, and acrid can become sweet, as is evident in grapes. But never can the plants be changed as a result of changes of their qualities. Therefore, nothing by which something can be brought about different from an animal or plant can be added to plants except sensible and transitive motion. And nothing can exist in the world except animals, vegetables, and minerals.

Robert Grosseteste was born probably shortly after Marius' treatise was written. While he was growing up and attending school, the number of Arabic treatises available in Latin was becoming ever larger, and Grosseteste was greatly influenced by them. As a young man, he accepted astrological doctrines, and even taught them. His treatise *On Prognostication* represents his classroom teaching at Oxford, probably prior to 1209. It presents the main outlines of astrological teaching much more fully and precisely than did Daniel of Morley's work, which was descriptive rather than technical. First he gives the natures of the twelve signs of the zodiac, then the nature of each of the planets. Then follows an explanation of the 'testimonies' and how they are determined by the *domus, exaltatio, triplicitas, terminus,* and *facies.* Each of these terms is explained, and then the significance of the five aspects (*oppositus, quartus, trinus, sextilis,* and *coniunctus*) is made clear. Further refinements resulting from the epicyclic motions of the planets are then thoroughly investigated and explained and are illustrated by the influence of the moon on the tides. Then, stepping to his equivalent of a blackboard, Grosseteste drew for his students a large diagram of the universe, with the circles of the planets and the zodiac clearly marked "so that you might understand all the foregoing without labor or tedium," and proceeded to give an example of how the weather might be forecast for a specific date (April 15, 1249) and to explain how to predict when it would be hot, cold, wet, or dry by calculating when the appropriate planets would be in positions of dominance.

But as Grosseteste grew older, he developed increasing reservations about astrology, and in one scientific work after another he gradually abandoned most of its teachings. His reasons, briefly stated, were first that it was bad science, pretending to knowledge it could not have, and second that it was anti-Christian, since it denied the omnipotence of God, the dignity of the human condition, and the freedom of man's will.

Selections from Robert Grosseteste, *Hexaëmeron*, translated from Oxford MS Bodley, lat. th. c. 17.

Even if we should suppose that the constellations have an effect on the works of free will, on events which are called fortuitous, and on human character, it would nevertheless not be possible for astrologers to make judgments about these things. For judgments of the stars are made according to the position of the sky and specific places of the heavens and their aspects, and the parts of the planets, their 'houses' and exaltations, and almost innumerable other things all gathered together, and according to a specific spot on the earth, in a specific moment of some question asked about an act or event which is going to occur, or in the moment of birth of a child or exact moment of a given year. But it is impossible for these things to be known with this kind of certainty by the astronomical art and astronomical instruments, to divide the minutes so precisely that an astrologer can say of two babies born in the same house, or even in the same city, at exactly or very nearly the same time, what the differences are in the constellations of their nativities. This is competely clear to those who realize what degree of certainty is possible to astronomers through their instruments. For it is not possible for those, however skilled in astronomy and however skilled in its operations, to assign the difference to two children thus born in a sign a degree differing by one minute, or even a second or sixtieth of a second. Nor can they determine the same sort of information concerning the heavens and other things which they themselves say are necessary for their judgments. Furthermore, they cannot distribute to two children born thus different events, natures or voluntary acts, so that they might say that this one will be chaste, prudent, brave, and rich, the other debauched, stupid, weak, ugly, and poor.... For there is not yet sufficient certitude concerning the motions of the heavens to enable the astronomers to know precisely in what indivisible moment of the year it is with respect to some specific place, nor are the positions of the planets at a precise moment truly known. This is clear to anyone who has spent much time and effort studying the astronomical tables....

It is not true, nor should it be conceded except for the sake of argument, that the stars have any effect on free will, or on the character and voluntary acts of men. For the free choice of a rational mind is subject to nothing in the rational order of things except God alone, but is rather placed above all corporeal creatures. Therefore, since the agent is more noble than the patient, a corporeal nature cannot, by its action, impress passions on the freedom of the will.... Therefore, whoever posits the efficacy of the stars on free will subjects the nature of the rational soul and the dignity of the human condition to a corporeal nature. And these people are enemies of human nature, since they subject it to what is naturally subject to it and deny that it is the image of God.... They also blaspheme against God, because they take away from God His dignity when they posit the rational mind, which they concede to be the image of God, lower than bodies....

But perhaps some astrologers will say that stars have incorporeal living and rational spirits, and by their spirits they act on the spirits of men, and by their bodies they act on the bodies of men. But this assertion is completely in vain, because even if we concede to them what they rashly assert, nevertheless in no way would it be true that the spirit of a star were superior by nature to the spirit of a man, since man, through his spirit, is the image of the Trinity. The authority of scripture leads to the same conclusion when it says in Deuteronomy: "The sun and moon and all the heavenly bodies created by God to serve all the nations under the heaven." If they were created to serve man, it is more natural that they should be acted upon, commanded, and receive orders from man than, on the contrary, they should act upon, command, and make impressions on man. Joshua exemplified this when he ordered the sun to stand still. Furthermore, the nature of free will is that it has sufficient power to bring about its own effects, with the help only of divine grace. But perhaps the professors of this sort of vanity will still say that the heavens make many obvious impressions on human bodies, and when the body undergoes some change, the soul undergoes it too. For the body, as the physicians say, follows the soul in the actions of the soul, and the soul follows the body in things undergone by the body. That is, when the body undergoes something, the soul undergoes it too, not because the body acts on the soul, but because the soul moves itself proportionally to the motion of the body to which it is united.... Therefore, because the heavens cause changes in bodies, and when bodies are changed the souls are also altered, astrologers of this sort will say that it lies within their province to make judgments on all the motions and passions of the soul which it has as a result of being united to a body. But one should answer them that the human body is subject to two movers. It receives many passions and impressions from the heavens, and it also receives motions and impressions from the action of its own soul. And since the soul, because of its rational power, is subject immediately to God, it has the power of commanding inferior powers and is more powerful in affecting its own body than are the heavenly bodies. ...

If someone should say that evil men who follow their lusts and carnal passions are subject to the judgment of the stars, one should answer that he who is now evil can suddenly be made good. Nor is it in the power of astrologers to say this, since the conversion of a man takes place through the operation of divine grace. ...

We wish finally to give this warning, that astrologers of this sort are seduced and seducers, and their teaching is impious and profane, written at the dictation of the devil, and therefore their books ought to be burned. And not only they, but those who consult them, are lost.

Several of Grosseteste's followers were unwilling to go so far. A Dominican writer, Richard Fishacre, borrowed extensively from the work of Grosseteste's we have just quoted but hedged on the matter of magic. Fishacre accepts the ability of astrologers and magi to foretell the future, and the reality and power of the demons who aid them. He only insists that such beings are evil and that it is wrong to worship them as gods.

Selections from Richard Fishacre, *Commentary on the Sentences*, translated from Balliol College MS 17.

Magi, astrologers, and prophets all predict the future. The magi predict what has been revealed to them by demons. The astrologers reveal what they have learned from the natural course of the stars, and some people believe that the three magi who worshipped Christ were really astrologers. Prophets foretell what has been revealed to them by God. ...

Hence it is clear that they err most seriously who consider either the magi or demons to be some sort of gods because they reveal hidden things and make such wonderful things occur that they are given the worship and sanctification that should be given to God alone. For just as we do not give divine honors to those perverse men who are skilled in the science of the stars just because they can reveal many hidden things to us, neither should we do more for the demons.

By Grosseteste's time, astrology had a wide enough following that it had to be taken seriously. His contemporary, Michael Scot, best known as a translator from the Arabic, was a renowned astrologer and was employed in this capacity by the emperor Frederick II. He was widely known and highly respected as a scholar in his lifetime, enjoying the patronage of the pope as well as the emperor. His mind was active, inquisitive, and retentive but not profound. He was an avowed Christian, who insisted that there was no incompatibility between his faith and "philosophy." Included in the term philosophy were the "sciences," many of them magical, and it was for his skill and learning in these that Michael was famous. Nothing at all is known about the time or place of his birth or about his education. But in 1217 he was in Toledo and had just finished his translation of Alpetragius' work on astronomy, and he was dead by 1235. His translating activity was extensive and of the first importance, and he had acquired a more thorough exposure to Arabic writings than most of his contemporaries. This undoubtedly influenced his world view to some extent. His whole outlook was saturated with magical notions, and although he occasionally denounces some aspects of magic as evil, he just as often forgets his condemnation and praises their usefulness. He seems to have known every folk superstition in Europe, and in addition he was an assiduous reader of books on magic. He is an excellent example of an intellectual trying to rationalize superstition and make it respectable, but what he accomplished was to make the natural universe magical.

A naturalistically conceived universe is one in which each part behaves consistently in accordance with its nature, and the nature in turn is determined by the physical constitution of the body. The treatise on the elements of the anonymous author and of Marius and the works of Grosseteste are good examples of this. A magical universe is one in which the parts have powers irrelevant to their natures and can bring about effects outside the course of nature.

And so objects, numbers, words, glances, and days, as well as the planets and constellations, have great power to do things which is not connected with their status as natural or rational beings. Michael accepted the natural universe, but he imputed to every part of it certain magical qualities.

The highest and most reliable guide to the secrets of the universe and to the future was astrology, which was useful in virtually every human activity, from determining the outcome of a war or whether one would recover from an illness to what the best day was to get a haircut and whether an unborn child would be a boy or a girl. Michael's major work, aside from his translations, was a comprehensive introductory work on the sciences, composed of a *Prooemium*, an introduction to the subject (*Liber introductorius*), a book concerning specific questions and problems (*Liber particularis*), and a book on physiognomy (*Physiognomia*). His descriptions of the characters of people born under the various planets is much the same as Daniel of Morley's, only much longer, and Michael considers the moon (as was more common among astrologers) to be a bad rather than a good planet, producing people who are poor and weak, unable to retain what wealth they have, lazy, fickle, morose, credulous, and stupid—much the same as Daniel's mercurial person. In any case, it was a very powerful planet, and depending on what sign it was in and whether it was waxing or waning, it either made people more hot, lively, and vigorous, quicker to learn and more disposed to sexual intercourse, or the reverse. Scot also goes into considerable detail about the constellations and their powers, which are at least as important as the planets in determining human character and the course of the life of one born under them. In addition to this sort of question, astrology is most useful in helping people make choices and in determining good and bad times to begin any sort of undertaking. Michael seems to have spent a good deal of his time making just this sort of judgment.

Michael was also well versed in medicine, an art for which he considered astrology essential. A competent physician must be skilled in astrology and consult the stars carefully before beginning any treatment. To determine whether a person will recover or die from an illness, Michael advises the physician to look at the ascendent and where the Lord is, since these both signify the patient and his condition. Also examine the moon for the patient, because it is a witness to the fortune of the sick man concerning whom the question was asked. Next see the tenth house and its Lord, since health and medicine are signified by them—also the power of the medicine and the advice of the physician. Then look at the sixth house and its Lord, by which the illness is denoted. Then see the eighth house and its Lord, by which death is indicated for the sick person. Then see the fourth house and its Lord, by which the end for each of the aforesaid things is truly indicated. For if the Lord of the ascendent is well disposed in the ascendent or the tenth or eleventh houses, it signifies that the patient will recover. ... But if the Lord of the ascendent is in the sixth or seventh house, it signifies a long siege of illness, and the physician is of little help. ... And if the Lord of the ascendent is in the eighth house, or the Lord of the eighth house is in the ascendent, it signifies death. And if the Lord of the ascendent is in the twelfth house, it signifies death, mourning, and burial, or that this illness cannot be cured by this physician. [Translated from Michael Scot, *Liber introductorius*, Paris, Bibliothèque nationale MS n. acq. lat. 1401].

In the case of certain psychological afflictions, the physician, if his medicine does not work, should send the patient to diviners or enchantresses, whose art is true even though it may seem an evil thing and contrary to the Christian faith.

Concerning alchemy, Michael seems to have been of two minds (this is not unusual with him). He clearly had a certain contempt for the alchemists themselves, and he apparently disbelieved in the possibility of transmuting the base metals into gold or silver. But of the art itself he expresses a much higher opinion in his descriptions of each of the magical arts in his *Prooemium*: "Alchemy, as it were, transcends the sky, because by the power of the four spirits it tries to transmute the base metals into gold and silver and to make from them a perfect water." [Translated from the same MS.] And if Professor Lynn Thorndike is correct in attributing a treatise on alchemy to Scot, he believed at one time that transmutation was possible, since he gives a procedure for making gold from copper.

"Take the blood of a ruddy man and the blood of a red owl, burning saffron, Roman vitriol, resin well pounded, natural alum, Roman alum, sugared alum, alum of Castile, red tartar, marcasite, golden alum of Tunis which is red, and salt." These ingredients are to be pounded in a mortar, passed through sieves, treated with the urine of an animal called *taxo,* or with the juice of wild cucumber, then dried, brayed again, and put into a crucible with the copper. [Reprinted by permission from Lynn Thorndike, *Michael Scot.* London: Thomas Nelson & Sons Limited, 1965, 113].

One of the most fascinating, controversial, and misunderstood thinkers of the Middle Ages was Roger Bacon. Bacon largely succeeded where Scot had failed in assimilating magic to nature. By and large, Bacon's conception of the universe was naturalistic, although it still contained some magical elements. However, when he affirmed a view which we would consider to be magical, he managed to provide an explanation which could be considered naturalistic if one makes certain allowances for the differences between Bacon's world view and ours.

Bacon was born probably in 1214 and died in 1292. Thus his life spanned the greater part of the thirteenth century. He was connected with both of the most renowned universities of northern Europe, Oxford and Paris, and he was a member of the Franciscan Order, which, along with the Dominicans, was probably the most vital force in the intellectual, social, and religious life of the thirteenth century. Roger joined the Franciscans in middle life after having lectured on the Arts faculty at Paris and having studied the sciences as an independent investigator (he seems to have come from a family of some wealth). His personality was such that he made powerful enemies within the Order, and he was thus prevented from attaining the fame and status which he so ardently desired and to which his intellectual abilities entitled him. As a Franciscan, he did continue to write and teach, but he complained bitterly of being required to do his regular work in the convent and of his lack of money for books, drugs, and equipment for his experiments.

Bacon was alarmed to the point of desperation by what he considered to be the sure signs that the time of Antichrist was at hand —immoral wars, changes of government, moral decay, corruption in high places in church and state, the credulity and stupidity of the masses and even of most of the educated, and the irrelevance of much of the university curriculum. What was needed to combat Antichrist was science—not just a collection of separate bits of knowledge, but a comprehensive, synthetic science of which languages, mathematics, and optics were the most important ingredients. This would make it possible to know the true meaning, both literal and allegorical, of Holy Scripture, to dispel error by true knowledge, to prolong human life greatly, and to defeat armies by means of wonderful new weapons.

His great opportunity came when in 1265 Clement IV, who as a cardinal had encouraged Bacon, became pope. The new pope sent Bacon a letter specifically requesting that this invaluable information be sent to him as quickly as possible. The result of this request was a series of three works, the *Opus maius*, *Opus minus*, and *Opus tertium*, which together constitute the greater part of Bacon's writings. Since the pope understood that the great work had already been written, but in fact Roger was simply asking for support in writing such a book, the results were written in great haste, are repetitive, and uneven in quality. The major assertions, which appear over and over again in them are that: 1. the current wretched state of the world is clear evidence that Antichrist is about to appear; 2. a universal science will make it possible to predict the future, to emerge victorious, and to reform society and the church; 3. most Latin scholars are stupid, foolish, credulous, and ignorant, but he and a small group of like-minded scholars possess true science; 4. that experimental science is the very highest type, because nothing is really known until it is known by experience. To Bacon the role of experimental science was primarily to confirm knowledge gained from authority, revelation, or reason, but it could also lead to new knowledge. Most of Bacon's knowledge was derivative, gained from books or conversation. He seems to have read many alchemical books and to have observed alchemists at work, although he was certainly not one himself. He had done a large amount of experimental work on optics and vision, and it is only here that his scientific work has any independent value. He was a man of great vision, imagination, and enthusiasm, but except in optics his actual contributions to human knowledge of the sciences were slight.

Bacon accepted without question the hierarchical structure of the universe and the axiom that the higher, worthier (*digniora*) parts could act on the lower, but not the other way around. He was contemptuous in the extreme of the lore and practice of the "magicians," considering them all to be frauds, deceiving the ignorant by their tricks and hocus pocus. But, although he specifically denied that mere words had any power to summon demons or work any kind of operation, he accepted the fact that words, considered as instruments of a rational mind, intent and wholly concentrating on its object, could be efficacious in bringing about the result desired by that mind. The real agent in such a case would be the rational mind, which could legitimately control those parts of nature inferior to it; the words were simply the means employed to help bring this about. But, although he roundly condemned the grosser and more obviously superstitious parts of magic, he considered both alchemy and astrology to be among those "most beautiful and useful" sciences which were essential to any fully educated man.

Bacon took over from Grosseteste much of the "Metaphysics of Light," according to which Light is the form of three dimensional corporeity and is also the efficient cause of all local motion. It propagates itself by a series of pulses, analogous to wave motion, spherically in all directions from its source. This was termed "multiplication of species" by both men. The strength of its operation depends in the first place upon the angle at which the ray of light strikes the object acted upon (the patient), the more nearly perpendicular, the greater its force; and secondly upon the distance the light must travel. Thus a short, straight ray has a maximum effect. Grosseteste had said repeatedly that the heavenly bodies act on lower bodies only by means of their light, and in his hands this was a strictly mechanistic physical theory. Bacon took over all of this and expanded it. He went beyond Grosseteste in investigating experimentally the properties of reflection and refraction and agreed with him that because light behaves according to definite geometrical laws, there could be no knowledge of nature without mathematics. He also went beyond Grosseteste in the role he assigned to light as an operative force in nature and in the kind of "species" which could be propagated in this way. Not only did the sun, moon, stars, and constellations affect the lower world by their light, but the human eye and voice, drugs and stones of certain types worked their powers on the things around them in the same way.

Bacon wrote one entire book *On the Multiplication of Species*, but he also summarized his teachings in many places throughout his writings. The following selections are taken from the *Opus maius*.

Selections from the *Opus maius*, translated from the Latin text of J. H. Bridges, ed., *The 'Opus Maius' of Roger Bacon*, 2 vols., (Oxford, 1897), I, 119–120, 395–398.

Now that we have considered these things about the multiplication of species, there are other things which must be considered concerning its further action. For light, by way of its multiplication, makes a luminous species, and this action is called univocal [that is, understood in a single sense] because the effect is univocal and of one kind and conforming to the agent. But there is also an equivocal [understood in more than one sense] multiplication, such as when light generates heat, heat generates putrefaction, putrefaction generates death, wine causes drunkeness, and thus with every agent which causes many effects extraneous to its own species and the force univocal to itself. And thus the sun and stars do all sorts of things here below, and the angels move the sky and the stars, and the soul moves its body. Nevertheless, the power of the agent does all these things, and this is the complete action of the agent and its power finally desired by nature. ...

And since the work of the rational mind is done especially and most efficaciously by means of words and formed intentions, an astrologer can form words at times which are chosen to have an ineffable power. For when the intention, desire, and power of the rational mind, which is more worthy than the stars, come together with the power of the sky, it is necessary that either a word or some other work be produced of amazing power to alter the things of this world so that not only natural things, but also souls, are inclined to that which the wise operator wishes (saving the freedom of the will), because the mind can follow celestial forces freely without compulsion, a subject which I have discussed in its own place.

From this root, the use of "characters" and incantations began among wise men. For characters are like images, and incantations are words brought about by the purpose of a rational mind, receiving the force of the sky at their very pronunciation. ... By this power dangerous animals are put to flight, some kinds of wild animals are summoned to your hand, snakes are called forth from caves and fish from the depths of the water. For the matter of the world is altered in many marvelous ways if these things are properly employed, and therefore they are very powerful against evil men and enemies of the commonwealth. But magicians of evil repute have brought the greatest infamy upon this science not only because they have abused the characters and incantations written by wise men to combat evil and to be highly useful, but they have added to them false incantations and vain and fraudulent characters, by which men are seduced. ... But all these things are of evil repute and outside the paths of the philosophers—indeed, they are against their teachings. And through these men the power of philosophy is defamed. And the theologians of our time, and Gratian, and many more holy men have reproved these useful and magnificent sciences along with the magical arts, not noting the difference between magic and the truth of philosophy. ...

I return therefore to the words and acts of the wise, formed by the force of the stars and the power of the rational mind. For just as a newborn child, exposed for the first time to the air and the world, receives the impressions of heavenly power from which he receives a basic complexion which he can never lose, ... thus it is with all newly-made things, which receive the force of the sky at the moment they come into existence; and that which they receive at the beginning they never lose, unless they are deprived of their natural being. And therefore, in these images, incantations, and characters, if they are made in a particular constellation, they receive and retain the power of the stars so that they might be able through these to act on the things of this world. And when the constellation in which the things of this sort were composed recedes, they recede. And since the rational mind is of greater dignity than the stars, therefore just as the stars, and all things, exercise their power and species on things outside themselves. ... the rational mind, which is the most active substance among all things (after God and the angels) can exercise and does continually exercise its species and power on the body, of which it is the actuality, and on things outside itself. And this is especially true when, from the strong desire and sure intention and confidence about which I speak, they not only receive power from the sky, but also from the rational mind which is more noble, and on account of this they can have a great power of altering the things of this world.

And if it should be said that just as works of this sort receive the force of the sky, thus also all things which are in the same region of the earth also receive the same force in the time of composition of words and works of this sort, and thus they all—men, cattle, horses, and trees—ought to have these powers, because the celestial rays of the constellation reach all of them at the same time; it should be said that this objection is not valid because not all these things are on the same horizon. For each separate point on the earth's surface is the center of a different horizon toward which the cones of the diverse pyramids of celestial force come, so that they can produce plants of different kinds on each separate tiny spot of the earth and make twins born of the same mother different in complexion and character, in the way they employ sciences and languages, in every-day affairs, and all other things. And furthermore because other things already made before the composition of the image [or what-ever], although they exist along with it, had nevertheless received at the time of their origin their own basic influence in accordance with which they were constructed, and therefore the force of the sky at this particular time I am speaking of does not have a natural effect on things of this sort already made as it does on those works and words newly made....

And if it should be said that everything which receives a force from a celestial constellation at the time of its origin, since that basic complexion would remain in it along with the continuing power of the constellation, it will alter things outside itself and change them sensibly, and especially at the time of its creation; and that through the procession of time, through continual other powers of the heavens, its original power will be weakened little by little until it is lost; it should be said that this is true, and depending upon the origins of such things, great changes sometimes occur, al-though we do not consider the source of these changes as we do in the case of comets or some other things.

Bacon's remarks on alchemy are obscure. We learn from them

that there are two kinds of alchemy, speculative and practical. Speculative alchemy seems to be much like primitive theoretical chemistry with a slight admixture of astrology; practical alchemy is much like industrial chemistry. Bacon considers them to be the "roots" of natural philosophy, theology, and medicine but complains that most people are ignorant of these essential sciences. He clearly knows a good deal about the arts and crafts which he connects with practical alchemy, and only in passing does he refer to gold making, although he considered it possible. The picture we get of the alchemist from Bacon's remarks is not that of a wizard in a peaked hat trying innumerable ways to transmute the baser metals into gold and discover the elixer of life, but rather of practical craftsmen—apothecaries, dyers, metal workers, goldsmiths, jewelers—trying to further their knowledge of their art. We also learn of the secrecy of the alchemical art—Bacon hardly dares write it to the pope, and even then he has written in code ("in an enigma") and has scattered his information throughout his various writings. The selection given below is actually the clearest statement he makes. If indeed he was communicating great secrets in code, his code was excellent. Nothing he says in any of his works reveals nearly as much as was (or is) easily available in many other alchemical books.

Selections from Roger Bacon *Opus tertium*, translated from the Latin text of J. S. Brewer, ed., *Fr. Rogeri Bacon Opera quaedam hactenus inedita*, I (London, 1859), 39–43.

There is another science, which concerns the generation of things from the elements and all inanimate things; about the elements and the simple and composite humors; about common stones, gems, and marbles; about gold and the other metals; about sulphur and the salts and pigments; about blue and red and other colors; about oils and burning coals and infinite other things concerning which we have nothing from the books of Aristotle; nor do either natural philosophers nor the whole crowd of Latins know anything about them. And because this science is unknown to all those who study natural philosophy, it must also be unknown to those who learn from them. But the generation of men and animals and vegetables from the elements and humors—things which they do study—has

much in common with the generation of inanimate things. Therefore, because of the ignorance of this science, ordinary natural philosophy cannot be known either, nor can speculative medicine, nor
consequently practical; not only because natural philosophy and
speculative medicine are necessary for practical medicine, but also
because all simple medicines made from inanimate things are known
by the science I am speaking of, as is clear in the second book of
Avicenna's *Canon medicinae*, where he enumerates the simple medicines. It is also clear from other authors, who do not know the names
or effects of these medicines except through this science. And this
science is speculative alchemy, which speculates on all inanimate
things and all generation of things from the elements.

There is also operative or practical alchemy, which teaches how
to make noble metals and many other things better and more abun-
dantly by art than they are made by nature. And this kind of science
is greater than any I have spoken of before because it produces
things of greater usefulness. For not only can it give wealth and
infinite other things to the commonwealth, but it teaches how to
find things that will prolong human life to a great age. We die sooner
than we should because we do not observe good health habits when
we are young, and because we inherited weak constitutions from
our fathers who did not take care of themselves either. And so old
age and death overtake us before the limit set by God has been
reached. Therefore this science is useful on its own account, but it
nevertheless certifies speculative alchemy through its works, and
therefore it certifies natural philosophy and medicine. This is evident
from the medical books, whose authors teach how to sublimate,
distill and resolve their medicines according to the operations of this
science, as is clear in healing waters, oils, and infinite other things.
Whence Galen, in his book *Dinamidiarum* [not really by Galen]
teaches physicians how to make *calcecuminon*, because the physicians, just as they don't know how to make it also don't even know
its name. And Avicenna in Book I of the *Canon medicinae* teaches
how to prove through the works of alchemy that not only blood, as
Galen thought, but also the humors nourish the body. But no physician knows or understands or does what these books teach. Therefore, this double science of alchemy is unknown to almost everyone.

For although many labor throughout the world to make metals and colors and other things, nevertheless very few know how to make colors truly and usefully; and almost no one knows how to make metals, and fewer still know how to do those things which will prolong life. There are also few who know well how to distill, sublimate, calcinate, resolve, and perform the other works of this art through which all inanimate things are certified and through which are certified speculative alchemy, natural philosophy, and medicine.

Next: There are not three people among the Latins who give themselves to this so that they might know speculative alchemy except insofar as it can be known without the works of practical alchemy—that is, who prove by experiments what the books and authors teach. There is only one man among all these who is completely able and skilled in this. And because so few know these things, they do not deign to share them with others, but only with each other, because they consider other men to be insane asses who bring petty arguments against them and who defile philosophy, medicine, and theology. It is very difficult and expensive to pursue these sciences, and so even those who know the art well are prevented from doing their work. And the books of this science are so obscure that a man can hardly find them, although there are more books on this subject than on any other. I have placed the roots of speculative alchemy in my discussion of Avicenna in the *Opus minus*. There in the discussion of the complete generation of things from the elements, I have also tried to certify with great diligence whatever should be known there about the paths of alchemy, natural philosophy, and medicine. These roots ought to be applied to the generation of all things. And I have explained this application in gold and other metals with great care. I have not proceeded further because my argument in that work did not require more. And my judgment is this, that when I touch upon these roots with their application to metals, they must, more than anyone thinks, be known about all the natural bodies which now exist, since without these roots, they study in vain about the branches, flowers, and fruits. I go too far here in words, but not in my heart, because I say this out of my grief over infinite error, and so that I might inspire you to a consideration of truth. Indeed, that man about whom I have spoken above [Aristotle], who composed so many and such great books on natural science, was ignorant of these foundations, and therefore his edifice cannot stand.

Much about the roots of practical alchemy is known through these things which I have written here. Nevertheless, I have phrased these things according to the words of the philosophers, and especially of Avicenna in the *Great Alchemy*, which he calls "The Book on the Soul Written in an Enigma" [not by Avicenna]. And I have placed these roots in the *Opus minus* as I indicated in the *Opus maius*, but in this third work I have presented them more exquisitely. But what I have said in the other places cannot be understood without what I say here nor can what I say here be understood without them. Nor do even all these things suffice for complete understanding except to those who are most wise and completely versed in this science; and there are not three such people in the world. For God has always concealed the power of this science from the multitude, for the crowd is not only ignorant of how to use these most worthy things but even converts them to evil purposes. Not even most philosophers have ever been able to attain these things, but only the wisest and most completely expert. I was unwilling to put the roots of these two sciences in the *Opus maius* because I did not then intend to write about them. But afterwards in the *Opus minus* I saw it to be appropriate, and I wrote those things which seemed to me to be expedient.

We could go on through the works of the next three centuries placing pro- and anti-magic texts against each other. St. Bonaventure, Minister General of the Franciscan Order during much of the time Bacon was a friar, was outspokenly opposed to astrology, for example. But we have seen enough now to get a reasonable idea of the range of opinion concerning magic, astrology, and alchemy, and how these terms were understood in the Middle Ages. During the twelfth century we find both uncompromising naturalism and credulous acceptance of magic ideas. During the early thirteenth century, the contemporaries Michael Scot and Robert Grosseteste were at opposite poles on the question. Later in the century, Roger Bacon seems to represent a minority view on the usefulness of astrology, alchemy, and the "philosophical" magic arts, and even he tried to make them conform to the prevailing naturalism of his time. Although late in the fourteenth century, there were still those (notably Nicole Oresme) who spoke out strongly against astrology, still during the fourteenth and fifteenth centuries, the number of alchemical and astrological works increased and their acceptance became much more general. Medical astrology was completely triumphant. The church itself, the bastion of opposition to magic lore, was deeply infected, and even popes consulted astrologers.

In the early sixteenth century, the Italian artist, Benvenuto Cellini, told of an amazing experience he had had in the Roman Colosseum, when he was transported by demons so that he could be with a girl he coveted. It is doubtful that any of his contemporaries doubted the literal possibility of such a story, whether they believed Cellini or not. But by now the Middle Ages were over.

CHAPTER 9

CONCLUSIONS

Despite the fact that many excellent illuminating studies of medieval science, as well as the texts of the works themselves, have been published in easily accessible volumes during the past fifty years, it is not unusual to find even well-educated people abysmally ignorant of the subject. Unfortunately this does not inhibit them from writing authoritatively about it. As recently as 1950, a Cambridge professor, M. Postan, contributed a chapter entitled "Why Was Science Backward in the Middle Ages?" to a book called *A Short History of Science* (Garden City, N.Y.: Doubleday and Co. Inc., 1959). While admitting some "purely intellectual and technical" progress, Prof. Postan asserted that compared to the Greeks and the moderns "all these achievements are bound to appear very poor." This poverty he attributed to a lack of scientific incentives, either intellectual or practical. Since the Middle Ages were an "age of faith," which "found the calls of faith itself... a task sufficient to absorb them, ... they had no time for occupations like science." And "even if there had been enough men to engage in activities as mundane as science, there would still be very little reason for them to do so.... Did not medieval men already possess in God, in the story of Creation and in the doctrine of the Omnipotent Will, a complete explanation of how the world came about and of how, by what means and to what purpose, it was being conducted? Why build up in laborious and painstaking mosaic a design which was already there from the outset, clear and visible to all?"

Sir William C. Dampier, author of one of the most widely used histories of science, *A History of Science and Its Relations with Philosophy and Religion* (4th ed., Cambridge: Cambridge Univ. Press, 1952), clearly felt bound by the rules of fair play to say something favorable about the Middle Ages, yet he seems to have found the task difficult, and the compliment he paid them is left-handed at best. Stating that scientific thought was "quite foreign to the prevailing mental outlook," which was enmeshed in a "tangle of astrology, alchemy, magic and theosophy," and completely hostile to original experimental investigation, he nevertheless admits that "their rational intellectualism kept alive, indeed intensified, the spirit of logical analysis, while their assumption that God and the world are understandable by man implanted in the best minds of Western Europe the invaluable if unconscious belief in the regularity and uniformity of nature, without which scientific research would never be attempted. As soon as they had thrown off the shackles of scholastic authority, the men of the Renaissance used the lessons which scholastic method had taught them. They began observing in the faith that nature was consistent and intelligible, and, when they had framed hypotheses by induction to explain their observations, they deduced by logical reasoning consequences which could be tested by experiment. Scholasticism had trained them to destroy itself" (p. 96). Thus Dampier attributes to the Renaissance one of the major achievements of the Middle Ages, and his views have largely dominated the modern textbook tradition.

Another widely read author, the philosopher Alfred North Whitehead, in *Science in the Modern World*, first published in 1925 and reprinted many times since, expresses much the same view but in starker terms. In order to contrast the modern world with the Middle Ages, he insists on "the inflexible rationality of medieval thought" as opposed to the scientific revolt of the seventeenth century, which he characterizes as an anti-intellectualist movement insisting on the observation of "irreducible and stubborn facts."

A diametrically opposite view of medieval science is expressed by William J. Brandt in *The Shape of Medieval History* (New Haven and London: Yale Univ. Press, 1966). He claims "that it was precisely a concern for practical problems, based upon close observation of details, that was the medieval contribution to the history of science." The reason medieval scientists made so little progress, he suggests (and he still takes this for granted), is that their perceptual framework made it impossible for them to relate these facts in a "scientific" way. "The thinkers of the twelfth and thirteenth centuries," he writes, "were defeated by their unconscious presuppositions. They visualized the physical universe as a field of objects, of discrete points; they attributed the activity and change visible on every side in the universe to forces, or virtues, located initially within one or another of these objects." Hence he concluded that medieval scientists lacked the modern notion of causality, since the unique causes of unique events inhered in objects and required no other explanation. This is an interesting and original interpretation, but it does not seem to describe very well what was actually done by medieval scientists.

Among scholars who actually know something about medieval science, the most debated question is no longer whether there was such a thing, but rather the relationship between it and modern science. The most prominent advocate of continuity is A. C. Crombie, who, in his *Robert Grosseteste and the Origins of Experimental Science, 1100–1700* (Oxford: Clarendon Press, 1953), presented a beautifully argued and well-documented case. His thesis is that what particularly distinguishes modern science from that of the Greeks is "its conception of how to relate a theory to the observed facts it explained, the set of logical procedures ... for constructing theories and submitting them to experimental tests," and that this is the creation of the Middle Ages. According to Crombie, "the strategic act by which Grosseteste and his thirteenth and fourteenth century successors created modern experimental science was to unite the experimental habit of the practical arts with the rationalism of the twelfth-century philosophy. Grosseteste appears to have been the first medieval writer to recognize and deal with the two fundamental methodological problems of induction and experimental verification and falsification which arose when the Greek conception of geometrical demonstration was applied to the world of experience. He appears to have been the first to set out a systematic and coherent theory of experimental investigation and rational explanation by which the Greek geometrical method was turned into modern experimental science. As far as is known, he and his successors were the first to use and exemplify such a theory in the details of original research into concrete problems. They themselves believed that they were creating a new science and in particular that they were creating a new methodology. Much of the experimental work of the thirteenth and fourteenth centuries was in fact carried out simply to illustrate this theory of experimental science, and all their writings show this methodological tinge." He then goes on to make clear what Grosseteste's methodology was (summarized in Chapter

3 above) and to trace its continuity through Roger Bacon, Theodoric of Freiberg, the Merton School, the fourteenth century Parisians, to northern Italy and Galileo.

The very heart of Crombie's thesis was attacked by the late Alexandre Koyré in his essay "The Origins of Modern Science: A New Interpretation," *Diogenes*, XVI (1956), 1–22. In the first place, Koyré denies the central importance of methodology in the history of science. "The place of methodology is not at the beginning of scientific development," he says, "but, we might say, in the middle of it. No science has ever started with a *Treatise on Method* and progressed by the application of such an abstractly devised method. ... Moreover, even if we admitted the prevailing influence of methodology upon scientific development, we would be faced with the paradox of seeing an essentially Aristotelian methodology producing, after a hundred years of sterility, a fundamentally anti-Aristotelian science." He also questions whether Grosseteste's method was in fact as Crombie presented it, and he strongly doubts its influence on later scientists.

On another level, Koyré attacks Crombie's assertion that the positivism of medieval science was an important ingredient in modern science. "Positivism," he says, "is a child of failure and renunciation. Its birthplace is Greek astronomy and its best expression is the Ptolemaic system. Positivism was conceived and developed not by the philosophers of the thirteenth century but by the Greek astronomers who, having devised and perfected the method and pattern of scientific thought (observation), found that they were unable to penetrate the mystery of the true movements of the heavenly bodies and who, therefore, restricted their aim to 'saving the phenomena,' that is, to the purely formal handling of observational data, a procedure which enabled them to make valid predictions, but which was paid for by the acceptance of a final divorce between mathematical theory and underlying reality.

"It was this conception—by no means progressive,... but utterly reactionary—that the positivists of the thirteenth century... tried to impose upon natural science. And it was in a revolt against this traditional defeatism that modern science, from Copernicus... and up to Galileo and Newton, accomplished its revolution against the shallow empiricism of the

Aristotelians, a revolution based upon a deep conviction that mathematics was much more than a mere formal devise for ordering data; in fact, the very key for the understanding of Nature."

The debate is by no means over, but the continuing appearance of critical texts and specialized studies of medieval science is making certain things much clearer than they had been.

As to whether medieval science was essentially rational and non-experimental, or was too preoccupied with the actual data of perception to be able to reach a sufficiently high level of abstraction, it can only be said that this depends on which specific medieval scientist one is talking about. Mathematicians such as Bradwardine and Swineshead certainly did not control their investigations by observation. They proved by mathematics what must inevitably be the case. But this was because they were mathematicians, not because they were medieval. It would be a task of considerable difficulty, however, to explain away the experimental side of the work of Theodoric of Freiberg, Witelo, Grosseteste, Albert the Great, Roger Bacon and a host of others. While it would be false to pretend that the works of these men were not guided to some extent by *a priori* considerations, it would be even more false to claim that they steadfastly refused to take cognizance of "irreducible brute facts," or to ignore the often highly-developed experimental method they employed.

To look at the same question from the other end, it is true that no medieval philosopher went so far as Galileo in treating cases of physical bodies in ideal situations which could never be observed, such as perfectly round balls rolling down frictionless planes, but medieval scientific works are rampant with ideal hypothetical situations. Perhaps medieval scientists took the world of their experience more seriously than did Galileo, but they were quite capable of reaching a respectable level of abstraction and generality. The difference here is one of success in execution, not one of conception.

A second consideration is whether the Middle Ages improved appreciably on the material they inherited from Antiquity, or whether the medieval period was just a pause. Here the evidence is clear beyond any question. Certainly in mechanics, optics, the tides, and methodology, the men of the Middle Ages added significantly to what they had inherited from the ancient world. This is also true of practical astronomy, such as the invention of new observational instruments, the improvement of astronomical tables, and proposals for calendar reform. And if no notable advance in theoretical astronomy was made during the Middle Ages proper, at least the way was prepared for the developments of the sixteenth through the eighteenth centuries.

The question of continuity is somewhat more complicated, but again, in some areas at least, the evidence leaves little doubt, although in others the issue is far from clear. It would be difficult, for example, to demonstrate the influence of Theodoric of Freiberg on Descartes' theory of the rainbow, since Theodoric's work was largely neglected after its publication. But in mechanics and astronomy there can be no doubt on the question of continuity, and in the important matter of method this is even more true. The way from Grosseteste through the Merton School, the Parisian scientists, to eastern Europe, Padua and other north Italian cities to Galileo has been quite well illuminated.

A more fruitful approach might be to stop arguing about the matter of continuity and try to understand rather what basic mutation occurred in the European world view between the fourteenth and seventeenth centuries. For medieval science is not, after all, modern science. One can hardly help being struck by both the modernity and the strangeness of much of the thirteenth and fourteenth century scientific writers. They graphed variations of qualities by a system of rectangular coordinates and came within an ace of inventing analytical geometry; they employed infinitesimals in their study of motion and seemed to be on the verge of inventing calculus; they came so close to the modern concept of inertia that several scholars have claimed this for them; they gave irrefutable arguments for the possibility of the earth's daily rotation; and they worked out many of the details of modern scientific method. But they did not invent either analytical geometry or calculus; they stopped short of a clear concept of inertia; they did not after all accept the fact of the earth's rotation; and they were much less rigorous in their method—especially its experimental side—than any modern scientists.

It is clear then that the more enthusiastic medievalists, such as Duhem, claim too much, but it is equally true that even the moderates, such as Brandt and Koyré, admit too little. It does not detract from the genius or originality of Galileo that he had predecessors —he did, after all, succeed where they had failed (or at least stopped short). Nor does it diminish the revolutionary importance of Copernicus' *De revolutionibus* that earlier astronomers had realized the possibility of the earth's rotation and that he repeated some of their arguments. Continuity is not incompatible with development. Modern science is not a direct outgrowth of Antiquity, without reference to the Middle Ages. It is rather the child of medieval science.

BIBLIOGRAPHICAL ESSAY

This bibliography is designed to be helpful to those people who are using this book. Therefore, it is highly selective, restricted to topics covered in the text and, except for the indispensable works of Anneliese Maier, Pierre Duhem, and Alexandre Koyré, to works in English. Since all the books listed contain extensive bibliographies of their own, it should be no problem for the reader to expand the list of articles and books on any of these topics as far as he wishes.

The pioneer, whose works definitely established the respectability of medieval science, was Pierre Duhem. His *Le Système du monde: histoire des doctrines cosmologiques de Platon à Copernic* (10 vols.; Paris, 1913–1916, reprint 1954), *Les Origines de la statique* (2 vols.; Paris, 1905–1906), and *Études sur Léonard de Vinci* (3 vols.; Paris, 1906–1913), although somewhat extreme in their claims for medieval science, contain a great wealth of information and valuable references to manuscript works. Other works which cover the whole range of medieval science are Lynn Thorndike, *A History of Magic and Experimental Science* (8 vols.; New York, 1923–1958); George Sarton, *Introduction to the History of Science* (3 vols.; Baltimore, 1927–1948), a comprehensive handbook treatment of the subject; and A. C. Crombie, *Medieval and Early Modern Science* (2 vols.; Garden City, N.Y., 1959), which is the only adequate textbook treatment of medieval science. The sections on the Middle Ages in all general histories of science should be scrupulously avoided.

Although late Antiquity and the early Middle Ages was a dismal time for scientific investigation, it was a period of paramount importance for maintaining the continuity of the scientific tradition. The only book I know which is devoted exclusively to the science of this period, and only to a limited aspect of it at that, is Samuel Sambursky, *The Physical World of Late Antiquity* (New York, 1962). Otherwise, a good, brief treatment is to be found in Marshall Clagett, *Greek Science in Antiquity* (New York, 1955; paperback ed. New York, 1963). Fuller and more detailed, and containing an excellent bibliography, is William Stahl, *Roman Science* (Madison, Wisc., 1962). Treating many of the same writers from the standpoint of literary culture as a whole is M. L. W. Laistner, *Thought and Letters in Western Europe*, *A.D. 500–900* (2nd ed., London, 1957).

For the science of the tenth and eleventh centuries, one must turn to periodical articles and sections of books and other topics. J. M. Clark, *The Abbey of St. Gall* (Cambridge, 1926) contains some excellent material, as does Harriet Lattin, *The Peasant Boy Who Became Pope: Story of Gerbert* (New York, 1951), which is much better than its title. By the same author is an article, "Astronomy: Our View and Theirs," in Lynn

White, Jr., ed., "Symposium on the Tenth Century," *Medievalia et Humanistica*, IX (1955), 13–17. See also J. W. Thompson, "The Introduction of Arabic Science into Lorraine in the Tenth Century," *Isis*, XII (1929), 184–193 and Mary C. Welborn, "Lotharingia as a Center of Arabic and Scientific Influence in the Eleventh Century," *Isis*, XVI (1931), 188–199.

Beginning with the twelfth century the literature becomes more voluminous, so what follows is more highly selective. Despite its title, Charles H. Haskins, *Studies in the History of Mediaeval Science* (Cambridge, Mass., 1924; reprint, New York, 1960) is devoted mainly to the twelfth century and to Michael Scot in the early thirteenth. Studies of particular scientific centers of the century are: for Chartres, J. A. Clerval, *Les Écoles de Chartres au moyen âge* (Paris, 1895) and J. M. Parent, *Le Doctrine de la création dans l'école de Chartres* (Paris, 1938), both of which should be read with the corrective chapter on Chartres in R. W. Southern, *Medieval Humanism and Other Studies* (New York, 1970); for Salerno, Paul O. Kristeller, "The School of Salerno," *Bulletin of the History of Medicine*, XVII (1945), 138–194, reprinted in P. O. Kristeller, *Studies in Renaissance Thought and Letters* (Rome, 1956), 495–551, and Brian Lawn, *The Salernitan Questions* Oxford, 1963); for Hereford, Josiah C. Russell, "Hereford and Arabic Science in England about 1175–1200." *Isis*, XVIII (1932), 14–25.

For the science of matter—what we would call chemistry—Robert P. Multhauf, *The Origins of Chemistry* (New York, 1967) supercedes other histories of early chemistry. Also helpful are Theodore Silverstein, "*Elementatum*: Its Appearance among the Twelfth-Century Cosmogonists," *Mediaeval Studies*, XVI (1954), 156–162 and "Guillaume de Conches and the Elements: *Homiomeria* and *organica*," *Mediaeval Studies*, XXVI (1964), 163–167, and Richard C. Dales, "Anonymi *De elementis*: From a Twelfth-Century Collection of Scientific Works in British Museum MS Cotton Galba E. IV," *Isis*, LVI (1965), 174–189, "An Unnoticed Translation of the Chapter *De elementis* from Nemesius' *De natura hominis*," *Medievalia et Humanistica*, XVII (1966), 13–19, and "Marius *On the Elements* and the Twelfth-Century Science of Matter," *Viator*, III (1972).

On Robert Grosseteste and the importance of the medieval methodological works, the most important book is A. C. Crombie, *Robert Grosseteste and the Origins of Experimental Science, 1100–1700* (Oxford, 1952). A much shorter presentation is A. C. Crombie, "Grosseteste's Position in the History of Science," in D. A. Callus, ed., *Robert Grosseteste, Scholar and Bishop* (Oxford, 1955), 98–120. Filling out and somewhat correcting Crombie are Richard C. Dales, "Robert Grosseteste's Scientific Works," *Isis*, LII (1961), 381–402, "The Text of Grosseteste's Treatise on the Tides, with English Translation," *Isis*, LVII (1966), 455–474, and *Roberti Grosseteste Commentarius in VIII Libros Physic-*

orum Aristotelis (Boulder, Colo., 1963). A complete bibliography of books about Grosseteste is Servus Gieben, O. F. M. Cap., "Bibliographia universa Roberti Grosseteste ab an. 1473 ad an. 1969," *Collectanea Franciscana*, XXXIX (1969), 362–418. The best general account of medieval optics is contained in A. C. Crombie, *Robert Grosseteste and the Origins of Experimental Science, 1100–1700*. Excellent studies of more specialized topics are William A. Wallace, O. P., *The Scientific Methodology of Theodoric of Freiberg* (Fribourg, 1959); David C. Lindberg, ed. and tr., *John Pecham and the Science of Optics: "Perspectiva Communis"* (Madison, Wisc., 1970), "Roger Bacon's Theory of the Rainbow: Progress or Regress?" *Isis*, LVII (1966), 235–248, "Alhazen's Theory of Vision and its Reception in the West," *Isis*, LVIII (1968), 321–341, "The Theory of Pinhole Images From Antiquity to the Thirteenth Century," *Archive for History of Exact Sciences*, V (1968), 154–176, and "The Cause of Refraction in Medieval Optics," *British Journal for the History of Science*, IV (1968), 23–38; Bruce S. Eastwood, "Robert Grosseteste's Theory of the Rainbow," *Archives internationales d'histoire de science*, no. 77 (Dec. 1966), 313–332, "Medieval Empiricism: The Case of Grosseteste's Optics," *Speculum* XLIII (1968) 306–321, and "Grosseteste's 'Quantitative' Law of Refraction," *Journal of the History of Ideas*, XXVIII (1967), 403–414; and Carl B. Boyer, "The Theory of the Rainbow: Medieval Triumph and Failure," *Isis*, XLIX (1958), 378–390.

Some of the best and most important studies of medieval science have been largely concerned with local motion. The scholar who did the earliest, and therefore almost necessarily faulty, work on the subject was Pierre Duhem, in *Le Système du monde, Études sur Leonard de Vinci, Les Origines de la statique*, and "Un précurseur français de Copernic: Nicole Oresme (1377)," *Revue générale des sciences pures et appliquées,* XX (1909), 866–873. Another Frenchman, Alexandre Koyré, took extreme exception to many of Duhem's conclusions, especially concerning the "modern" character of medieval science, in the three fascicles of his superb *Études Galiléenes* (Paris, 1939). The basis for a sounder knowledge of the subject was first laid by Anneliese Maier in a series of painstaking studies based on extensive manuscript evidence: *Die Vorläufer Galileis im 14. Jahrhundert* (Rome, 1949), *Zwei Grundprobleme der scholastischen Naturphilosophie* (2nd ed.; Rome, 1951), *An der Grenze von Scholastik und Naturwissenschaft* (2nd ed.; Rome, 1952), *Metaphysische Hintergründe der spätscholastischen Naturphilosophie* (Rome, 1955), and *Zwischen Philosophie und Mechanik* (Rome, 1958). In 1952, E. A. Moody and Marshall Clagett collaborated on *The Medieval Science of Weights* (Madison, Wisc., 1952). Since then a series of critically edited texts, usually with English translation and commentary, has been published in the series Publications in Medieval Science by the University of Wisconsin Press: H. Lamar Crosby, Jr., ed. and tr., *Thomas of Bradwardine His Tractatus de Proportionibus* (1955), Curtis Wilson, *William Heytesbury: Medieval Logic and the Rise of Mathematical Physics* (1956), Marshall Clagett, *The Science of Mechanics in the Middle Ages* (1959), *Nicole Oresme and the Medieval Geometry of Qualities and Motions, A Treatise on the Uniformity and Difformity of Intensities Known as Tractatus de configurationibus qualitatum et motuum* (1968), Edward Grant, ed. and tr., *Nicole Oresme De proportionibus proportionum and Ad pauca respicientes* (1966), and *Nicole Oresme and the Kinematics of Circular Motion, Tractatus de commensurabilitate et incommensurabilitate motuum celi* (1971). Also valuable for this topic are James A. Weisheipl, O. P., *The Development of Physical Theory in the Middle Ages* (New York, 1959; reprint 1971), which is primarily interested in the conceptual problems, and Marshall Clagett's early work, *Giovanni Marliani and Late Medieval Physics* (New York, 1941).

Considering the great recent popularity of the occult sciences, there are very few competent general works on them. There is a great deal of material on both alchemy and astrology in Lynn Thorndike, *A History of Magic and Experimental Science*. For alchemy E. J. Holmyard, *Alchemy* (Baltimore, 1957) is an excellent though somewhat technical treatment. For a more purely descriptive and historical approach, see Robert Multhauf, *The Origins of Chemistry*. On astrology, in addition to Thorndike, Duhem's *Le Système du monde* contains a wealth of material. Three more specialized studies are Lynn Thorndike, *The Sphere of Sacrobosco and its Commentators* (Chicago, 1949) and *Michael Scot* (London, 1965), and Richard J. Lemay, *Abu Ma'shar and Latin Aristotelianism in the Twelfth Century* (Beirut, 1962). For medieval attitudes toward astrology, see Theodore O. Wedel, *The Medieval Attitude Toward Astrology, Particularly in England* (New Haven, 1920) and R. C. Dales, "Robert Grosseteste's Views on Astrology," *Mediaeval Studies*, XXIX (1967), 357–363.

On the question of the continuity between medieval and modern science, the earliest and most extreme exponent of the "continuity" thesis is Pierre Duhem. The most important and convincing recent work to stress this view, especially with regard to method, is A. C. Crombie, *Robert Grosseteste and the Origins of Experimental Science, 1100–1700*, although Crombie slightly tempered his thesis in "The Significance of Medieval Discussions of Scientific Method for the Scientific Revolution," in Marshall Clagett, ed., *Critical Problems in the History of Science* (Madison, Wisc., 1959), 79–101. Written before the medieval achievement in methodology was known, and finding a highly developed "modern" scientific method at the University of Padua (where Galileo taught) in the fifteenth century, is the epoch-making article by J. H. Randall, "The Development of Scientific Method in the School of

Padua," *Journal of the History of Ideas*, I (1940), 177–206. Making what seems to me to be an irrefutable case for Galileo's dependence on his medieval predecessors are two articles by E. A. Moody, "Galileo and Avempace: The Dynamics of the Leaning Tower Experiment," *Journal of the History of Ideas*, XII (1951), 163–193, 375–422 and "Galileo and his Precursors," in Carlo L. Golino, ed., *Galileo Reappraised* (Berkeley, 1966), 23–43. The major exponents of discontinuity are the extremely erudite studies by Alexandre Koyré, *Études Galiléenes* and *From the Closed World to the Infinite Universe* (Baltimore, 1957; paperback ed. New York, 1958), which convincingly argue the point that there was a revolutionary change in the European world view during the fifteenth and sixteenth centuries, without which the scientific revolution could not have occurred. More in the nature of a debating speech, and to my mind missing the point, is the same author's "The Origins of Modern Science: A New Interpretation," *Diogenes*, XVI (1956), 1–22. Also arguing for discontinuity are two essays by Ernan McMullin, "Medieval and Modern Science: Continuity or Discontinuity?" *International Philosophical Quarterly*, V (1965), 103–129 and "Empiricism and the Scientific Revolution," in Charles S. Singleton, ed., *Art, Science and History in the Renaissance* (Baltimore, 1968), 331–369. Abandoning argument in favor of trying to understand more clearly just what happened between the fourteenth and seventeenth centuries are Herbert Butterfield, *The Origins of Modern Science* (New York, 1951) and E. J. Dijksterhuis, *The Mechanization of the World Picture* (Oxford, 1961), both intelligent works of synthesis rather than bits of original research. The definitive book on this subject is yet to be written.

One of the main purposes of this book has been to make available to the student translations of actual works of medieval science which were not otherwise available. It is of necessity highly selective. But there are other English translations to which one can turn to supplement what is presented here. At the present time, Edward Grant, ed., *A Source Book in Medieval Science* is in press. When it appears, a very large proportion of all aspects of medieval science will be available in English. Meanwhile one may read William H. Stahl, tr., *Macrobius, Commentary on the Dream of Scipio* (New York, 1952), Ernest Brehaut, tr., [Isidore of Seville] *An Encyclopedist of the Dark Ages* (New York, 1912), William D. Sharp, M. D., tr., *Isidore of Seville, The Medical Writings* (Philadelphia, 1964), L. W. Jones, tr., *Cassiodorus Senator, An Introduction to Divine and Human Readings* (New York, 1946), George W. Corner, *Anatomical Texts of the Earlier Middle Ages* (Washington, 1927), Hermann Gollancz, tr., *Dodi ve Nechdi ... To Which is Added the First English Translation of Adelard of Bath's Natural Questions* (London, 1920) (this is a very unreliable translation), and A. E. Waite, tr., *The New Pearl of Great Price. A Treatise Concerning the Treasure and Most*

Precious Stone of the Philosophers (London, 1894). In addition to these titles, all the works in the University of Wisconsin's Publications in Medieval Science contain large amounts of translated material, some being in fact translations as well as editions, and the general works, such as Crombie, *Robert Grosseteste and the Origins of Experimental Science, 1100–1700* and Thorndike, *A History of Magic and Experimental Science* and *The Sphere of Sacrobosco* translate considerable sections of the works they are discussing or give English summaries of them.